PAST PAPERs of AMC 8 vol. I

by *Kay*

Table of Contents

01. Linear Equation

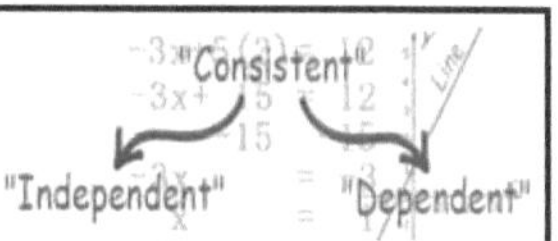

A. Solving Equation

① making expression and equation
② expand any brackets and collect like terms
③ use inverse operations to isolate the unknown and maintain balance

B. Solving Linear System

① substitution method
- solve one of the equations for one of its variables.
- substitute the expression into the other equation and solve for the other variable

② elimination method
- multiply one or both of the equations by a constant to obtain coefficients that differ only in sign for one of the variables.
- add the revised equations and combining like terms will eliminate one of the variables. Solve for the remaining variable.
- substitute the value into either of the original equations and solve for the other variable.

Exactly one solution

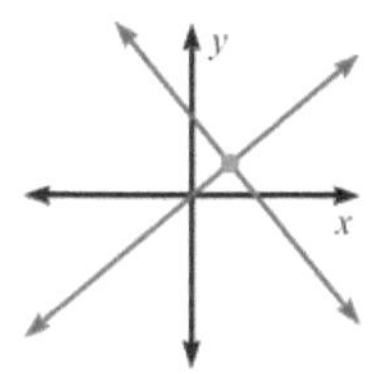

Infinitely many solutions

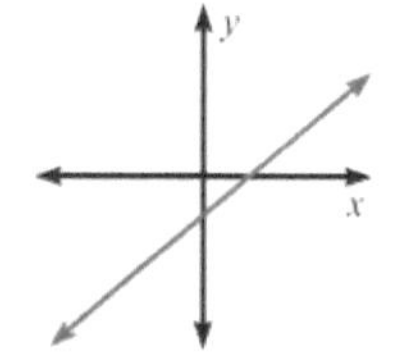

No solution

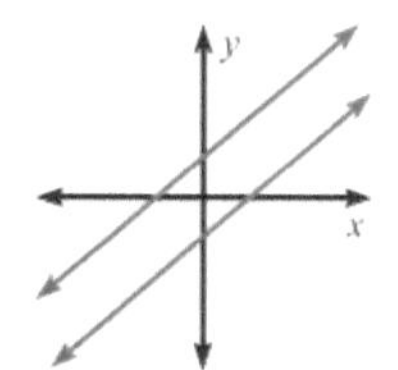

01.

Jessie went to the fruit store and bought 15 fruits for a total of $25. Tomatoes he bought cost $1 each, pineapples he bought cost $4 each, and watermelons he bought cost $5 each.

If she bought at least one of each type, how many tomatoes did Jessie buy?

(A) 6

(B) 7

(C) 10

(D) 12

(E) 13

02.

Ms. Aida needs to know the combined weight in kilograms of three boxes she wants to send a package by delivery service. However, the only available balance has some part of all balance weights so, it is not use for weights less than 5 kilograms or more than 10 kilograms. So the boxes are weighed in pairs in every possible way. The results are 7.5, 8.5 and 8.0 kilograms.

What is the combined weight in kilograms of the three boxes ?

(A) 11.5

(B) 12.0

(C) 12.5

(D) 13.0

(E) 13.5

03.

Beatrice wants to buy a 700 dollars iPad. For her birthday, her grandmother send her 80 dollars, her uncle sends her 70 dollars and her father gives her 100 dollars. She earns 50 dollars per week for her part-time job. She will use all of her birthday money and all of the money she earns from her part-time job.

In how many weeks will she be able to buy the iPad ?

(A) 7 weeks

(B) 8 weeks

(C) 9 weeks

(D) 10 weeks

(E) 11 weeks

04.

Martin bought a bag of oranges at the grocery store. He gave one-thirds of the oranges to Naomi. Then he gave Olivia 5 apples, keeping 3 apples for himself.

How many oranges did Martin buy at the grocery store ?

(A) 10

(B) 12

(C) 14

(D) 16

(E) 18

05.

Becky, Cassidy, and Dave played badminton with each other. Becky won 3 games and lost 1 games. Dave won 2 games and lost 4 games.

If Cassidy won 2 games, how many games did she lose ?

(A) 0

(B) 1

(C) 2

(D) 3

(E) 4

06.

Starting with some apples and some empty baskets, William tried to put 12 apples in each basket, but that left 3 baskets empty. So instead he put 8 apples in each basket, but then he had 4 apples left over.

How many apples did William have?

(A) 72

(B) 76

(C) 80

(D) 84

(E) 88

07.

It is assumed that $n\#m$ means $n - 2m$. What is the value of a, for

$$a * (3 * a) = 4$$

(A) -4

(B) -2

(C) 1

(D) 2

(E) 4

08.

Nathan's language arts homework is to read a novel in four weeks. On the first week, Nathan read $\frac{1}{6}$ of the pages plus 25 more, and on the second week he read $\frac{1}{5}$ of the remaining pages plus 32 pages. On the third week he read $\frac{1}{4}$ of the remaining pages plus 15 pages. He then realized that there were only 78 pages left to read, which he read the fourth week.

How many pages are in the novel ?

$(\boldsymbol{A})$ 208

$(\boldsymbol{B})$ 226

$(\boldsymbol{C})$ 240

$(\boldsymbol{D})$ 264

$(\boldsymbol{E})$ 282

09.

The Mableton Zoo has only two-legged birds and four-legged mammals. On one visit to the zoo, Gilbert counted 124 heads and 326 legs.

How many of the animals that Gilbert counted were four-legged mammals?

$(\boldsymbol{A})$ 39

$(\boldsymbol{B})$ 45

$(\boldsymbol{C})$ 52

$(\boldsymbol{D})$ 63

$(\boldsymbol{E})$ 72

10.

A group of students riding on bicycles and tricycles rode past Bonsu primary school. The school principal, Veronica, counted 8 students and 21 wheels.

How many bicycles were there?

$(\boldsymbol{A})$ 3

$(\boldsymbol{B})$ 4

$(\boldsymbol{C})$ 5

$(\boldsymbol{D})$ 6

$(\boldsymbol{E})$ 7

11.

In a science quiz with fifteen problems, a student gains three points for a correct answer and loses one point for an incorrect answer.

If Zinna answered every problem and his score was 25, how many incorrect answers did he have?

(A) 3

(B) 5

(C) 7

(D) 10

(E) 12

12.

Jessie went to the fruit store and bought 15 fruits for a total of \$25. Tomatoes he bought cost \$1 each, pineapples he bought cost \$4 each, and watermelons he bought cost \$5 each.

If she bought at least one of each type, how many tomatoes did Jessie buy?

(A) 6

(B) 7

(C) 10

(D) 12

(E) 13

13.

Six friends went to see a movie and agreed to share the movie ticket price equally. Because Thomas forgot his money, each of his five friends paid an extra $1.40 to cover his portion of the total ticket price.

What was the total movie ticket price?

(A) $30

(B) $36

(C) $38

(D) $40

(E) $42

14.

Using only pennies, nickels, dimes, and quarters, what is the sum of the largest number and the smallest number of coins Jasmine would need so she could pay any amount of money equal to 0.89 dollar?

$(\boldsymbol{A})$ 89

$(\boldsymbol{B})$ 97

$(\boldsymbol{C})$ 99

$(\boldsymbol{D})$ 103

$(\boldsymbol{E})$ 111

15.

Jeju International Middle School has the same number of 6th grade students and 7th grade students. Four-fifths of the 6th grade students and three-fourths of the 7th grade students went on a field trip to the museum.

What fraction of the students were 6th grade ?

$(A)\ \frac{1}{2}$

$(B)\ \frac{3}{4}$

$(C)\ \frac{16}{31}$

$(D)\ \frac{3}{5}$

$(E)\ \frac{17}{32}$

16.

On the last Halloween day evening, Mr. Samuel gave sweet lollipops to his relatives. He gave each boy as many lollipops as there were boys in the relatives. He gave each girl as many lollipops as there were girls in the relatives. He brought 50 lollipops, and when he finished, he had nine lollipops left. There were one less girls than boys in his relatives.

How many girls were in his relatives ?

(A) 3

(B) 4

(C) 5

(D) 6

(E) 7

02. Venn Diagram

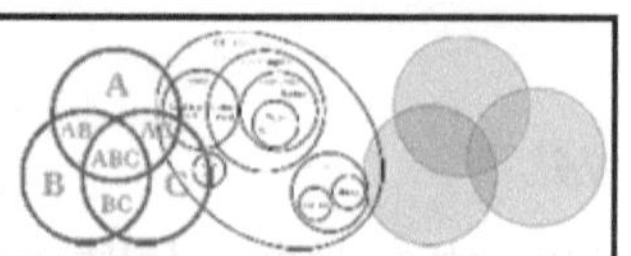

A. Set

① elements of a set are the objects or members which make up the set

② A is a proper subset of B if every element of A is also an element of B, but $A \neq B$. We write $A \subset B$.

③ universal set U contains all of the elements under consideration

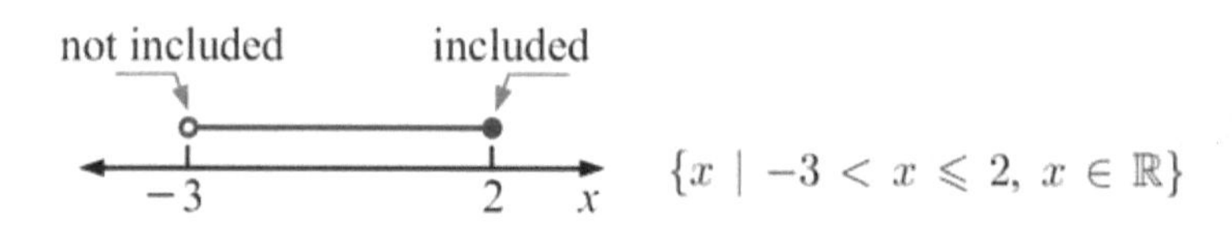

B. Venn Diagram

A Venn diagram consists of a universal set U represented by a rectangle, and sets within it that are generally represented by circles.

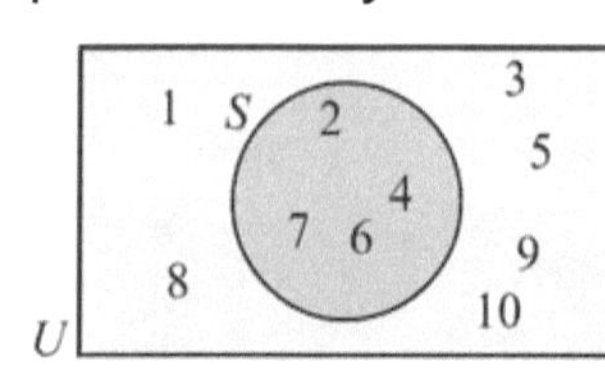

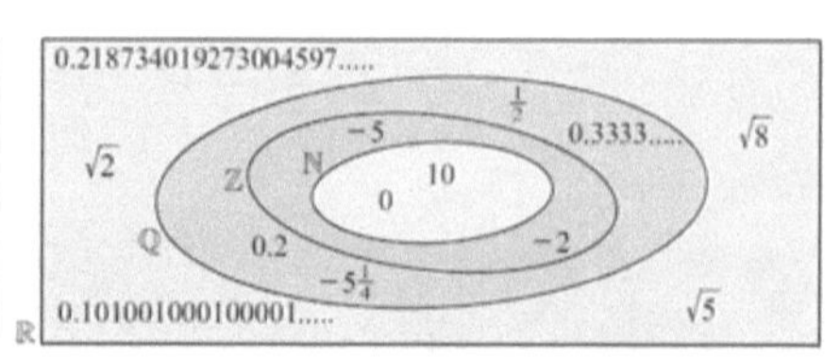

① union of two sets
This set contains all elements belonging to A or B or both A and B.

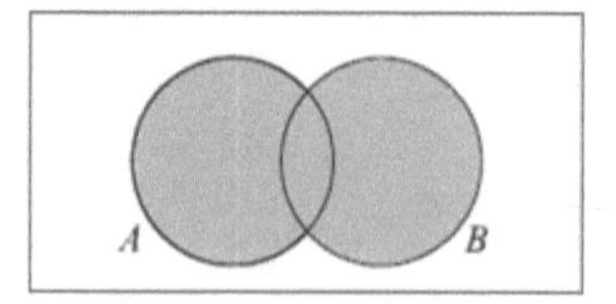

② intersection of two sets
This is the set of all elements common to both sets.
So,

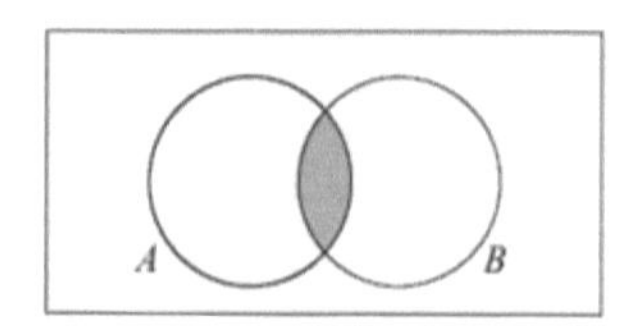

$$n(A \cup B) = n(A) + n(B) - n(A \cap B)$$

17.

At Jefferson City, 514 people voted on two issues in a local referendum with the following results: 325 voted in favor of the first issue, constructing waste disposal site, and 272 voted in favor of the second issue, building city park. Suppose that there were exactly 68 people who voted against both issues.

How many people voted in favor of both issues?

$(\boldsymbol{A})$ 151

$(\boldsymbol{B})$ 168

$(\boldsymbol{C})$ 174

$(\boldsymbol{D})$ 295

$(\boldsymbol{E})$ 446

18.

In a town of 405 families, every family owns a sedan, pickup truck, or both.

If 229 families own sedans and 218 families own pickup trucks, how many of the sedan owners do not own a pickup truck ?

(A) 42

(B) 176

(C) 187

(D) 209

(E) 219

19.

Sets X and Y have the same number of elements. The intersection set of set X and set Y has 512 elements and the union set of set X and set Y has 818 elements.

What is the number of elements in set Y ?

(A) 153

(B) 207

(C) 306

(D) 353

(E) 665

20.

How many sets of two elements that are different whole numbers can be added to the number set {0, 1, 2, 3, 4, 5, 6, 7, 8, 9, 10} so that the average of the remaining numbers is 5 ?

(A) 3

(B) 4

(C) 5

(D) 6

(E) 7

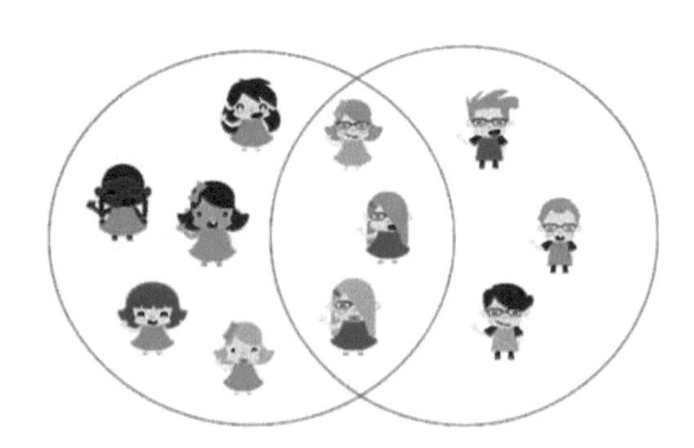

03. Pythagorean Theorem

A. Pythagorean Theorem

A right angled triangle is a triangle which has a right angle as one of its angles. The side opposite the right angle is called the hypotenuse and is the longest side of the triangle.

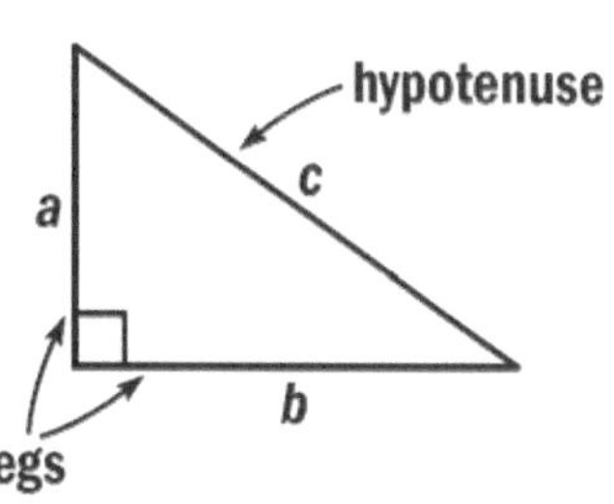

The other two sides are called the legs of the triangle.

For any right triangle, the sum of the squares of the lengths of the legs equals the square of the length of the hypotenuse. So,

$$c^2 = a^2 + b^2$$

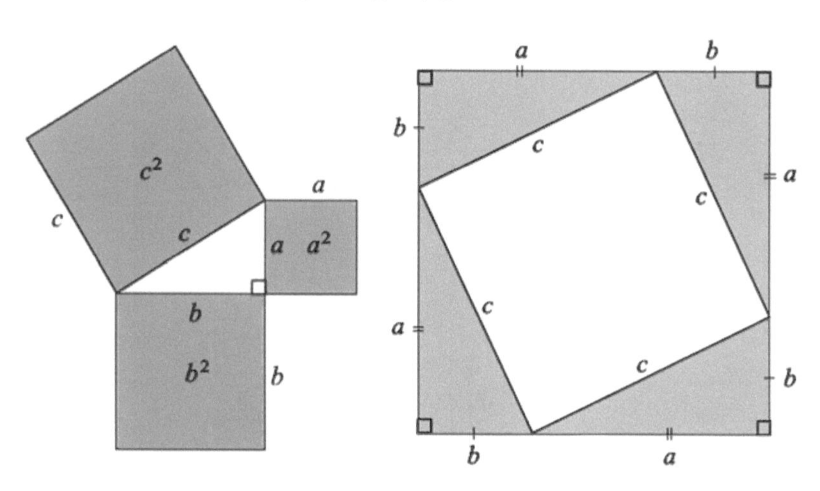

B. Pythagorean Triple

A Pythagorean triple consists of three positive integers a, b, and c, such that $a^2 + b^2 = c^2$, a well-known example is $(3, 4, 5)$. If (a, b, c) is a Pythagorean triple, then so is (ka, kb, kc) for any positive integer k.

21.

Aida rode $\frac{1}{3}$ mile west, then $\frac{2}{3}$ mile north, and finally $\frac{2}{3}$ mile west with her new bicycle.

How many miles was she, in a direct line, from her starting point?

(A) 1

(B) $\frac{4}{3}$

(C) $\frac{\sqrt{5}}{3}$

(D) $\frac{5}{3}$

(E) $\frac{\sqrt{13}}{3}$

22.

What is the perimeter of trapezoid $PQRS$?

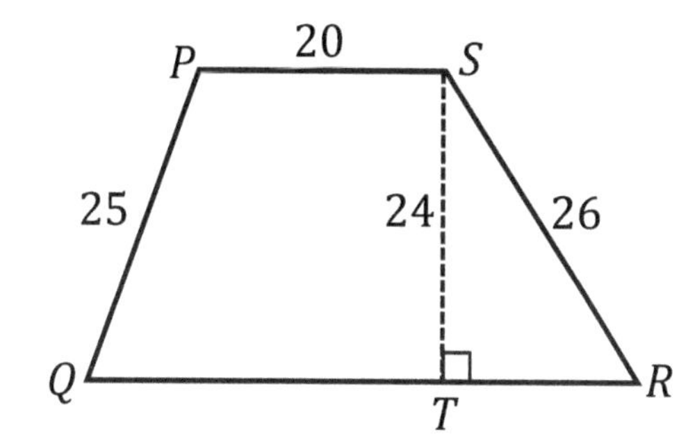

(A) 108

(B) $10\sqrt{7}$

(C) $10\sqrt{15}$

(D) 121

(E) $12\sqrt{7}$

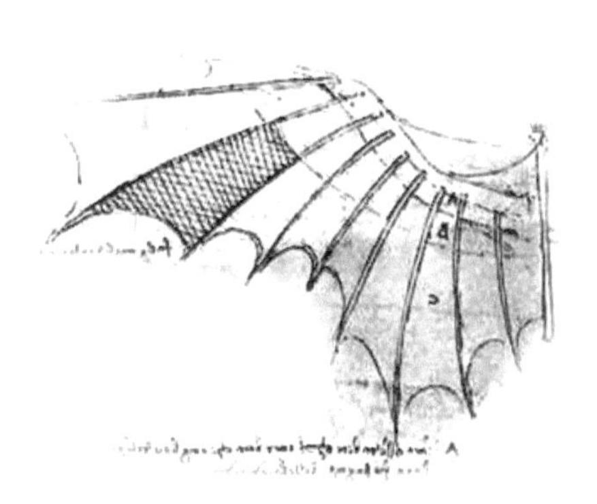

23.

Squares are constructed on the sides of a $5-12-13$ right triangle, as shown as right.
A capital letter represents the area of each square.

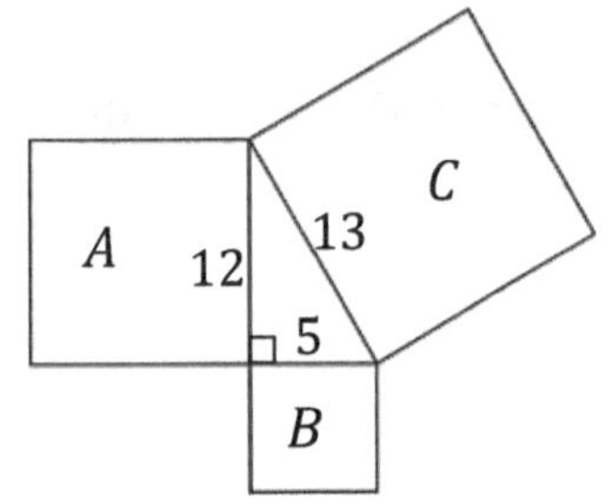

Which one of the following is correct expression ?

(*A*) $A = B + C$

(*B*) $A = C - B$

(*C*) $A + B = C - B$

(*D*) $2C = 2A + 2B$

(*E*) $C^2 = A^2 + B^2$

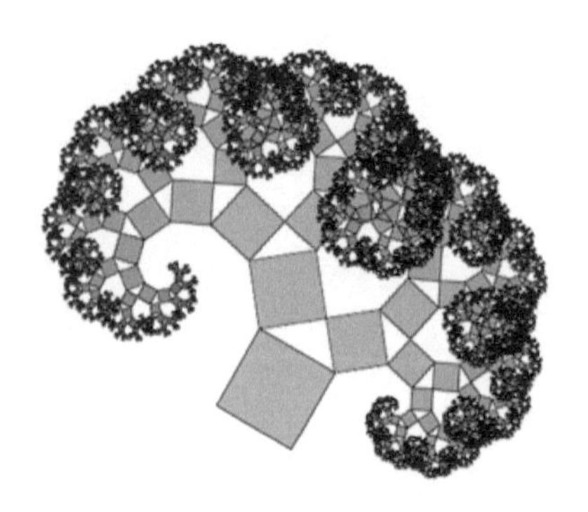

24.

In $\triangle OPR$, $\overline{OP} = \overline{OR} = 13$, and $\overline{PR} = 10$.

What is the area of $\triangle OPR$?

$(\boldsymbol{A})$ 30

$(\boldsymbol{B})$ 45

$(\boldsymbol{C})$ 60

$(\boldsymbol{D})$ 90

$(\boldsymbol{E})$ 120

25.

In figure below, a right triangle $\triangle OPQ$ and a rectangle $\square PQRS$ have the same area. They are joined to form a trapezoid, as shown.

What is the length of the side $\overline{OQ}$?

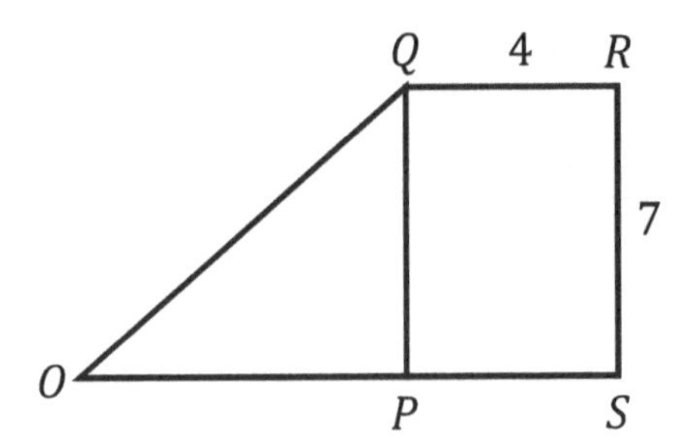

$(\boldsymbol{A})$ 3

$(\boldsymbol{B})$ $\sqrt{65}$

$(\boldsymbol{C})$ 11

$(\boldsymbol{D})$ $\sqrt{113}$

$(\boldsymbol{E})$ $2\sqrt{29}$

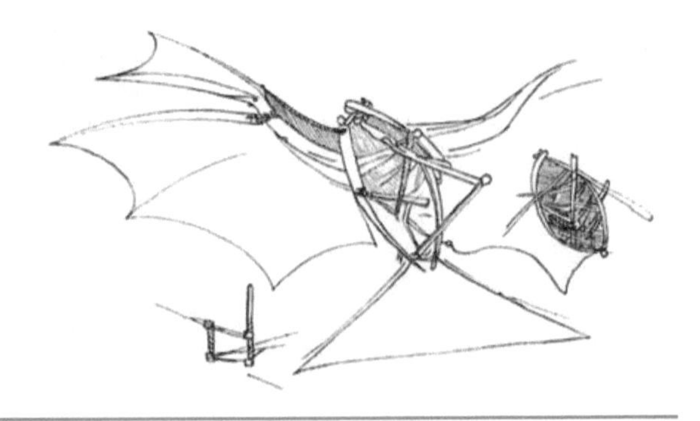

26.

Angle $\angle ABC$ of $\triangle ABC$ is a right angle. The sides of $\triangle ABC$ are the sides of squares as shown below. The area of the square on $\overline{AB}$ equals 64, and the perimeter of the square on $\overline{AC}$ has length 68.

What is the side length of the square on $\overline{BC}$?

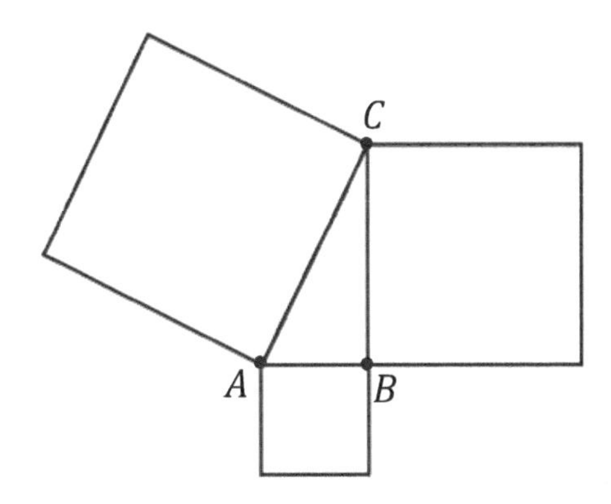

(A) 12

(B) 13

(C) 14

(D) 15

(E) 16

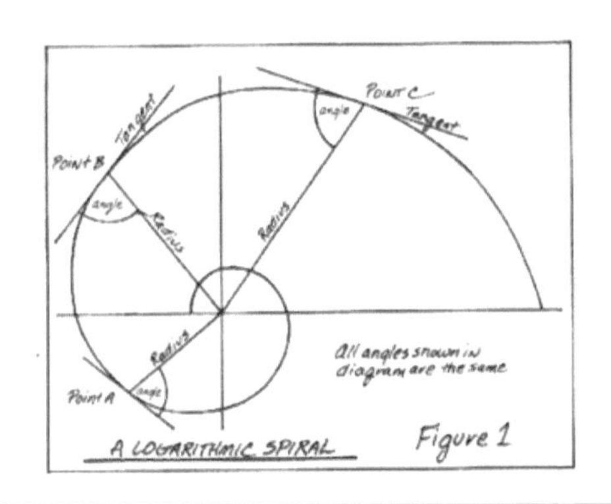

27.

In the figure below, a square is inscribed in a square with area 196, with one vertex of the smaller square on each side of the larger square. A vertex of the smaller square divides a side of the larger square into two segment at a ratio of $3:4$.

What is the area of the smaller square ?

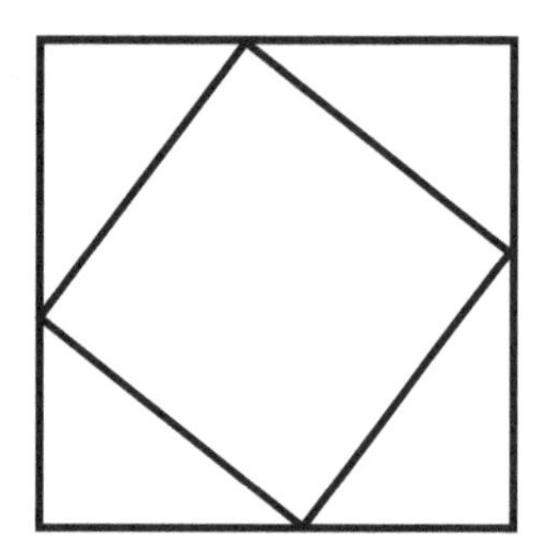

(A) 72

(B) 81

(C) 96

(D) 100

(E) 144

28.

Let X be the area of an isosceles triangle with sides of length $26, 26,$ and 20. Let Y be the area of another isosceles triangle with sides of length $26, 26,$ and 48.

What is the difference of $Y - X$?

(A) -28

(B) 0

(C) 28

(D) 46

(E) 50

29.

A unit octagram is composed of a octagon and its 8 congruent right triangular extensions, as shown in the figure below.

What is the difference of the area of the original octagon and the area of the extended triangles ?

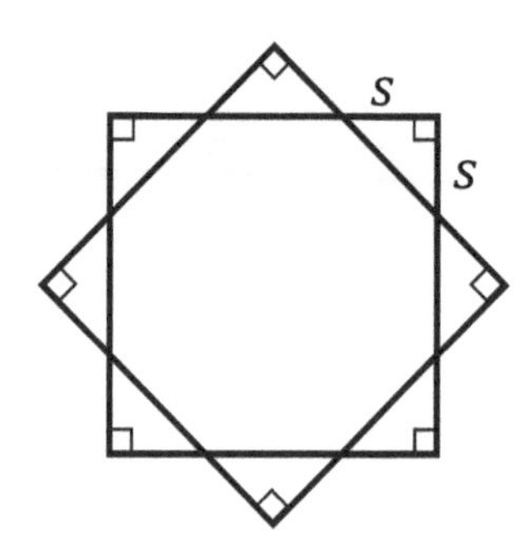

(*A*) $8s^2$

(*B*) $4\sqrt{2}s^2$

(*C*) $(8-\sqrt{2})s^2$

(*D*) $(2+4\sqrt{2})s^2$

(*E*) $(3+2\sqrt{2})s^2$

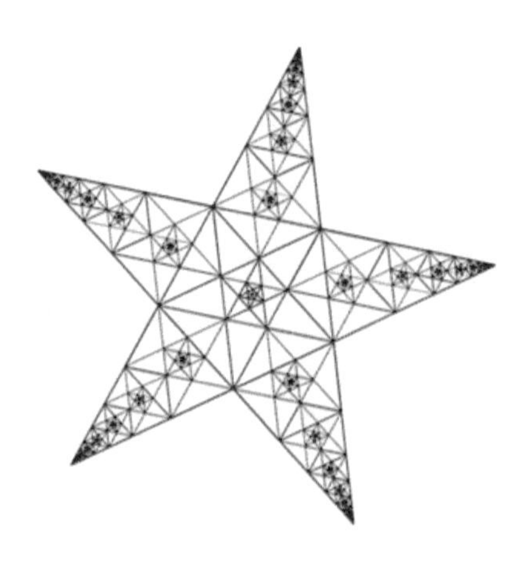

30.

The base side length of an isosceles triangle is 16 and its area is 48.

What is the perimeter of the isosceles triangle ?

(A) 36

(B) 38

(C) 40

(D) 42

(E) 44

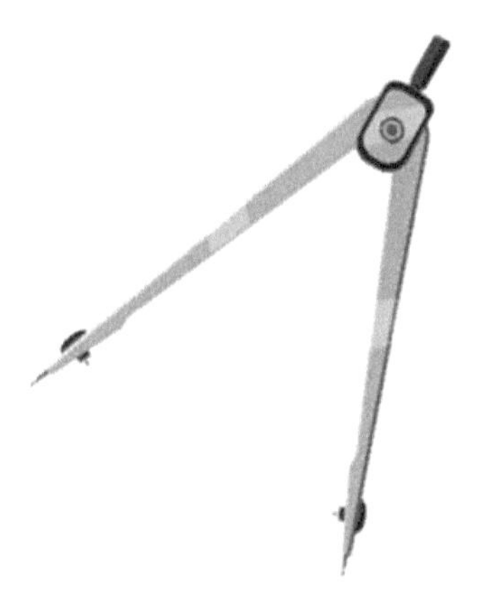

31.

In the figure right, the triangle $\triangle PQR$ is an isosceles right triangle with $\overline{PQ} = \overline{QR}$. The line segment $\overline{PX}$ is one of third of the line segment $\overline{PR}$ and $\overline{PS}$ is 6 units long.

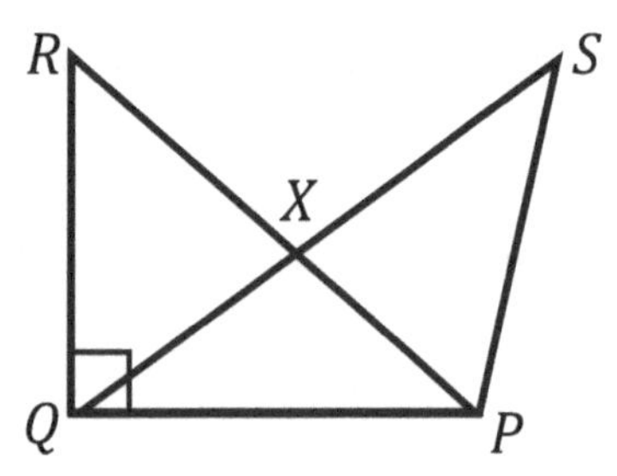

If the side length of $\overline{PS}$ is equal to $\overline{PQ}$, what is the length of $\overline{RX}$?

(*A*) $2\sqrt{2}$

(*B*) 3

(*C*) 4

(*D*) $4\sqrt{2}$

(*E*) $6\sqrt{2}$

32.

A semicircle encloses an isosceles triangle PQR of area 2. The circle has its center O on hypotenuse PR and the vertex P is placed at circumference of the semicircle.

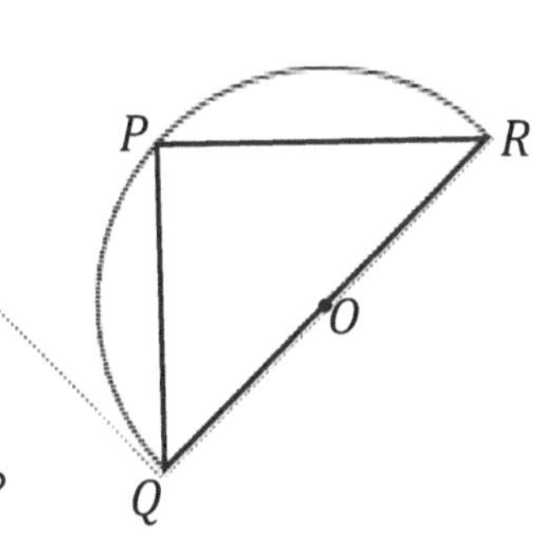

What is the area of the semicircle ?

(A) $\frac{\pi}{2}$

(B) π

(C) $\sqrt{2}\pi$

(D) $\frac{2}{\sqrt{2}}\pi$

(E) 2π

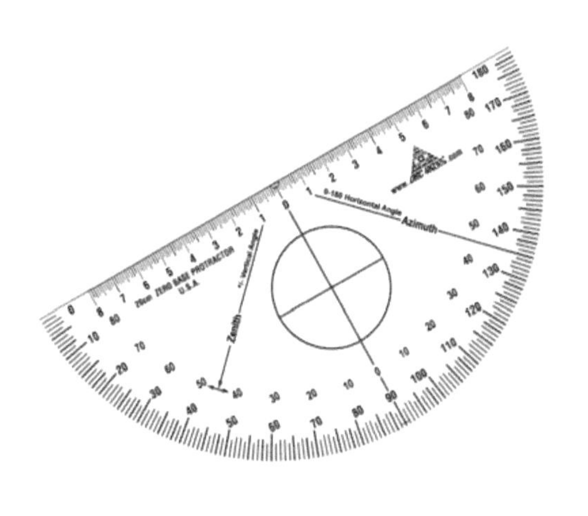

33.

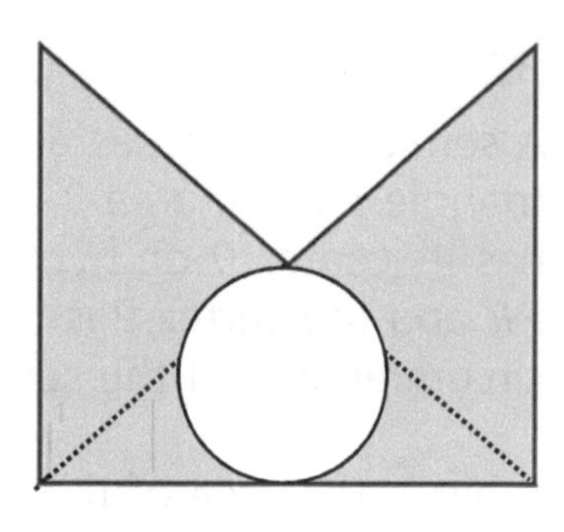

Two isosceles right triangle with leg length 4, intersect at right angles, bisecting their hypotenuse, as shown as right. The circle's diameter is the segment between the two points of intersection and base leg.

What is the area of the shaded region created by removing the circle from the two isosceles right triangles?

(*A*) $12 - 2\pi$

(*B*) $12 - \pi$

(*C*) $14 - 2\pi$

(*D*) $14 - \pi$

(*E*) $16 - 2\pi$

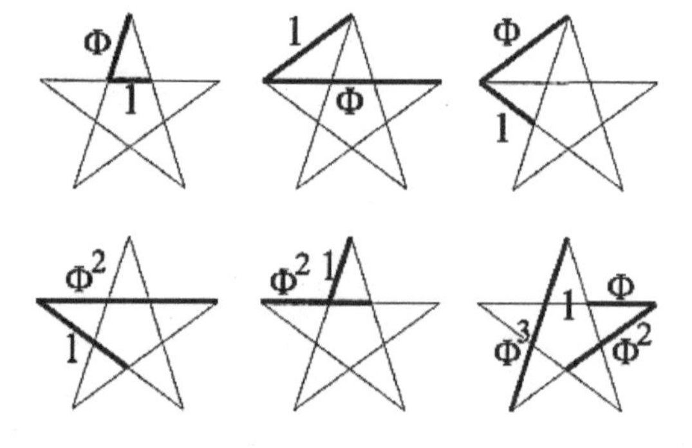

34.

In the figure right, choose point S on $\overline{PR}$ so that $\triangle PQS$ and $\triangle RQS$ have equal perimeters.

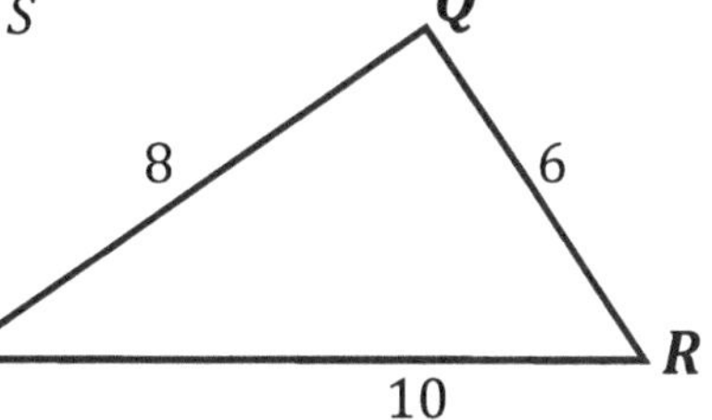

What is the area of $\triangle RQS$?

(A) 8.0

(B) 8.4

(C) 8.8

(D) 9.2

(E) 9.6

35.

In the concave quadrilateral $\square OPQR$ shown right, $\angle ORQ$ is a right angle, $\overline{OP} = 24, \overline{PQ} = 26$, $\overline{QR} = 8$, and $\overline{OR} = 6$.

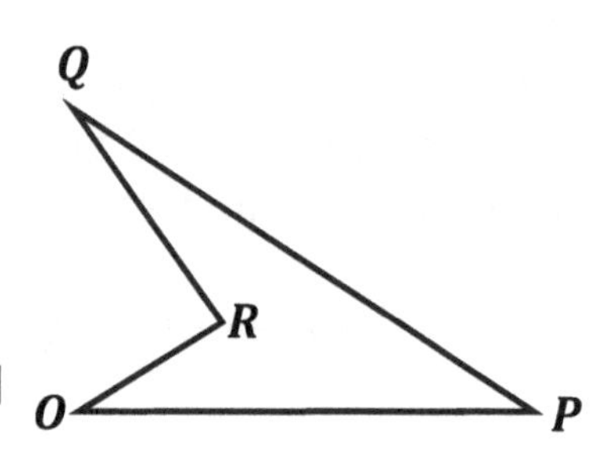

What is the area of quadrilateral $\square OPQR$?

(A) 88

(B) 90

(C) 96

(D) 102

(E) 108

36.

In the figure below, $\square OPQR$ is a trapezoid, $\overline{OP} = 40$, $\overline{PQ} = 30$, $\overline{OR} = 40$, and $\overline{ST} = 24$.

What is the area of the trapezoid?

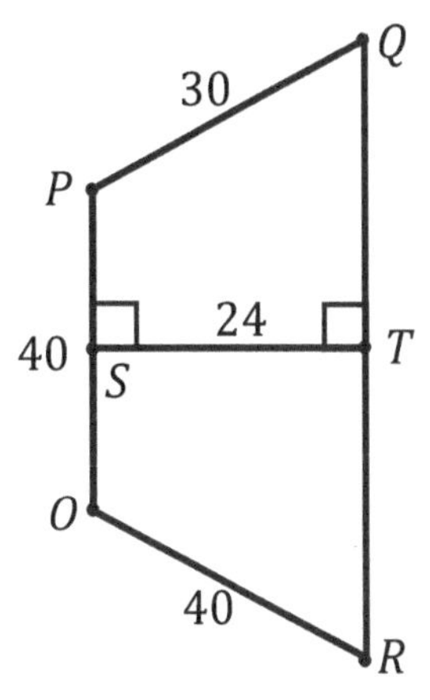

(A) 900

(B) 960

(C) 1,024

(D) 1,560

(E) 1,808

37.

In the figure below, the trapezoid $ABED$ is composed of, a square $ABCD$ and a right triangle CDE.

What is the area of the trapezoid $ABED$?

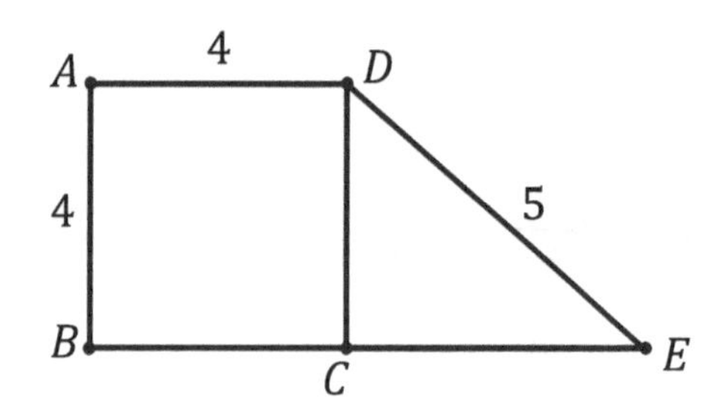

(A) 20

(B) 22

(C) 24

(D) 25

(E) 28

38.

In the figure right, $\square ABCD$ is a square and $\square XYWZ$ is a parallelogram.

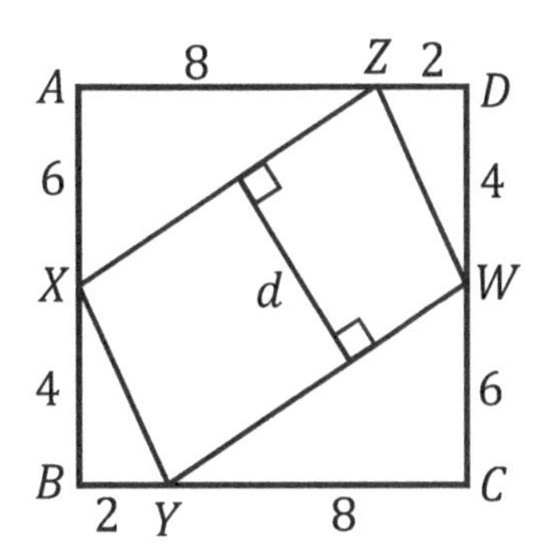

What is the length of d ?

$(\boldsymbol{A})$ 3.6

$(\boldsymbol{B})$ 4.0

$(\boldsymbol{C})$ 4.4

$(\boldsymbol{D})$ 5.2

$(\boldsymbol{E})$ 5.8

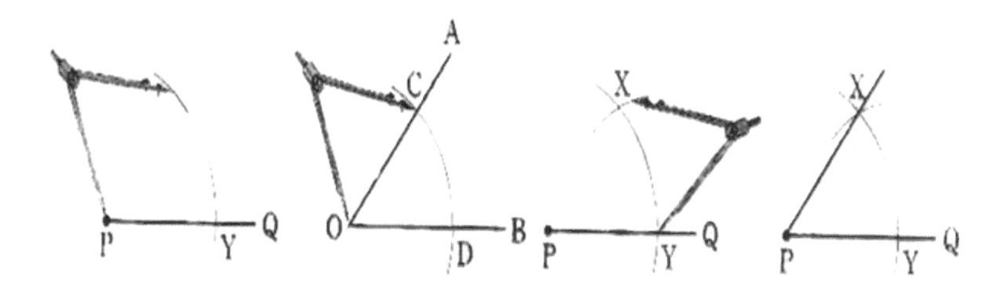

39.

The following below figures are composed of three congruent squares and different circles.

What is the sum of the areas of all shaded regions ?

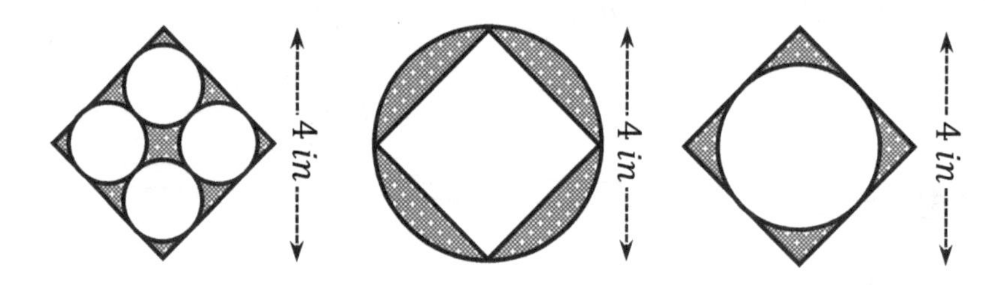

(A) 4

(B) 4π

(C) 8

(D) $12 + 4\pi$

(E) $24 - 4\pi$

40.

A 2 by 3 rectangle is inscribed in a semicircle with shorter side on the diameter.

What is the area of the semicircle?

(A) 3π

(B) 4π

(C) 5π

(D) 6π

(E) 7π

41.

A square is inscribed in a circle with radius 2 and circumscribed about another circle as shown below.

Which fraction is the ratio of the area of the square's shaded region to the area between the two circles ?

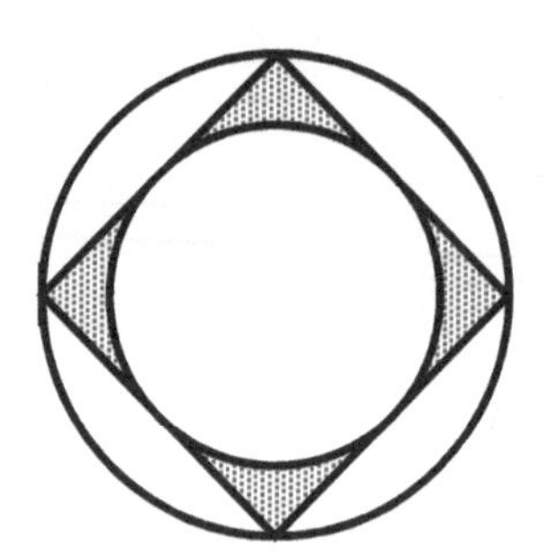

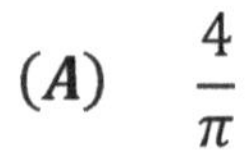

(A) $\frac{4}{\pi}$

(B) $\frac{\pi}{2}$

(C) $\frac{\pi + 4}{2}$

(D) $\frac{4 - \pi}{2\pi}$

(E) $\frac{4 - \pi}{\pi}$

42.

In the figure below, the two circles pictured have the same center O. The side $\overline{AC}$ of an equilateral triangle is tangent to the inner circle at B, $\overline{AC} = 16$, and the radius of smaller circle is 6.

What is the area of the shaded region ?

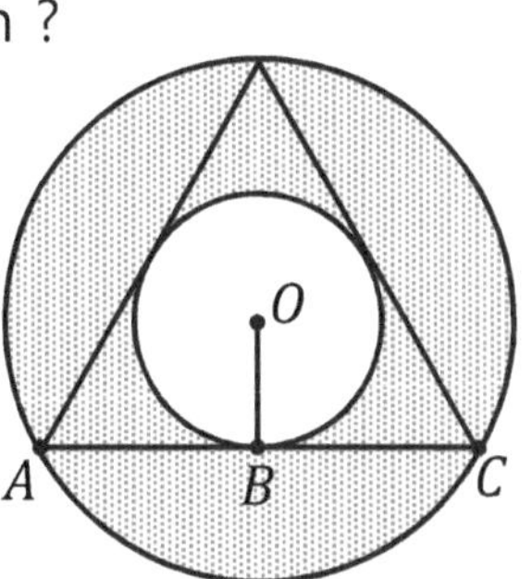

(**A**) 25π

(**B**) 36π

(**C**) 49π

(**D**) 64π

(**E**) 81π

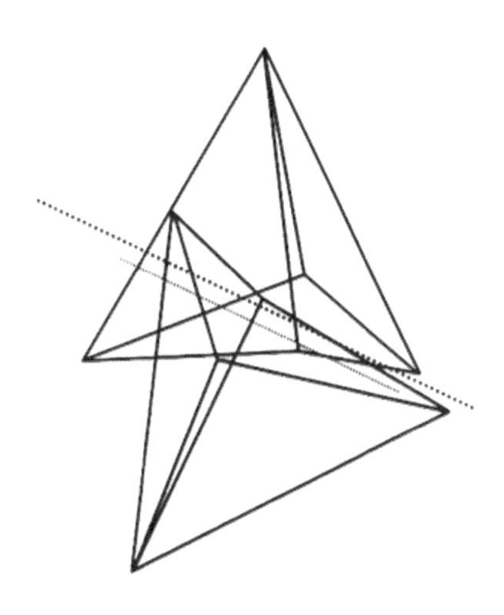

43.

In the standard (x, y) coordinate plane, the two semicircles AOD and BOC pass through the origin O. What is the ratio of the combined areas of the two semicircles to the area of the larger circle O ?

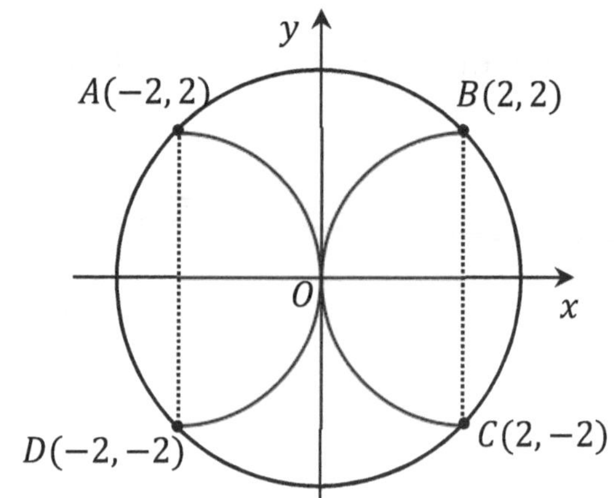

$(\boldsymbol{A})$ $1:4$

$(\boldsymbol{B})$ $1:2$

$(\boldsymbol{C})$ $2:3$

$(\boldsymbol{D})$ $3:4$

$(\boldsymbol{E})$ $5:8$

44.

Given the areas of the two semicircles in the right figure,

what is the area of the triangle ?

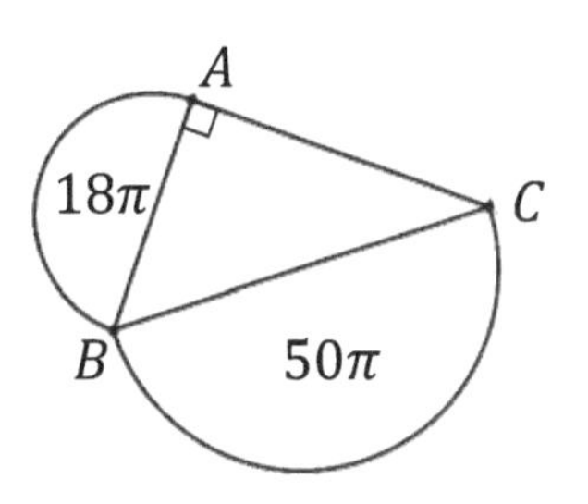

(***A***) 32

(***B***) 32π

(***C***) 40

(***D***) 40π

(***E***) 84

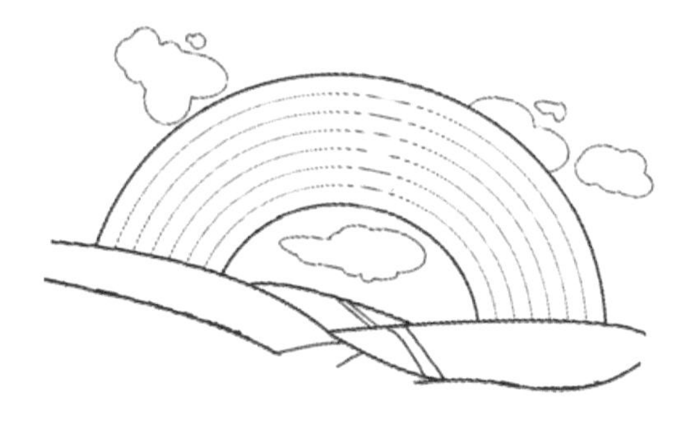

PRIMUS INTER PARES
K·DEAN

04. Prime Factorization

A. Prime Number

A prime number is a whole number that is greater than 1 and has exactly two whole number factors, 1 and itself. A composite number is a whole number that is greater than 1 and has more than two whole number factors. The number 1 is neither prime nor composite.

Examples of Prime and Composite Numbers		
Number	**Factors**	**Prime or composite?**
24	1, 2, 3, 4, 6, 8, 12, 24	Composite
41	1, 41	Prime
51	1, 3, 17, 51	Composite

B. Prime Factorization

When you write a number as a product of prime numbers, you are writing its prime factorization.
You can use a diagram called a factor tree to write the prime factorization of a number.

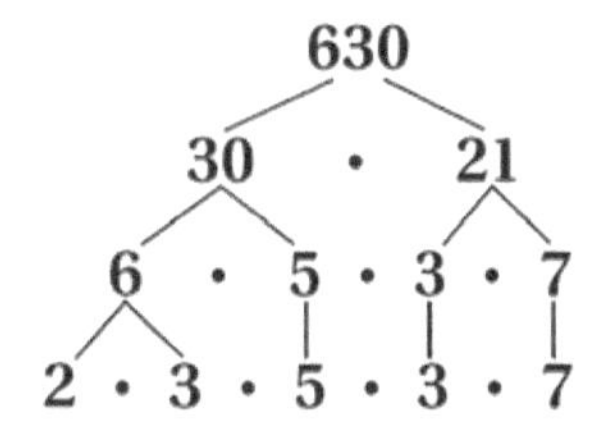
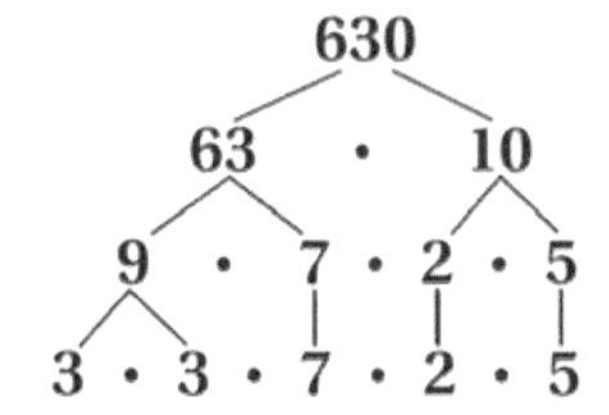
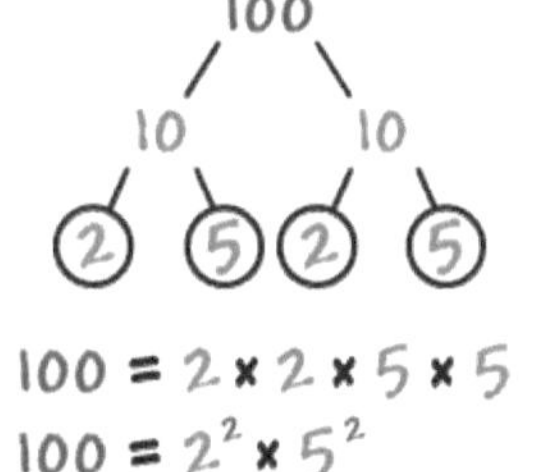
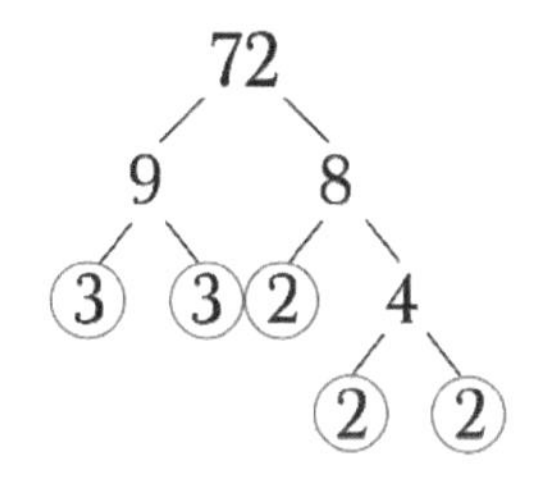

45.

The sum of two prime numbers is 115.

What is the difference of these two prime numbers ?

(A) 99

(B) 102

(C) 105

(D) 108

(E) 111

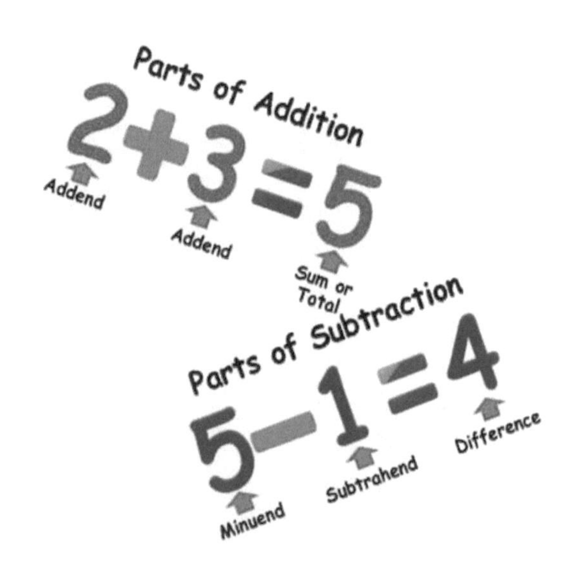

46.

Three members of the Pythagoras Middle School Math club had the following conversation.

Paul: I just realized that our math test scores are all two-digit prime numbers.

Richard: And the sum of your two scores is greater than 155.

Samuel: Oh really. The sum of your scores is less than 155.

Paul: And the difference of your two scores is 2.

What number is Paul's score ?

$(\boldsymbol{A})$ 67

$(\boldsymbol{B})$ 71

$(\boldsymbol{C})$ 73

$(\boldsymbol{D})$ 79

$(\boldsymbol{E})$ 83

47.

What is the smallest whole number that is neither prime number nor perfect square and that has prime factor greater than 60 ?

(A) 3599

(B) 3721

(C) 4209

(D) 4899

(E) 5183

48.

In the math class, students guess that the age of their math teacher, Kay, is $37, 38, 39, 42, 44, 45, 46, 47, 48$, and 49. Kay says, "At least half of you guessed too high, two of you are off by one, and my age is a prime number."

How old is Kay ?

(A) 35

(B) 37

(C) 39

(D) 41

(E) 43

49.

In how many ways can 400,004 be written as the sum of two prime numbers?

(A) 0

(B) 1

(C) 2

(D) 3

(E) 4

50.

Kenneth wrote 6 different numbers, one on each side of three cards, and laid the cards on a table, as shown right. The differences of the two numbers on each of the three cards are equal and the three numbers on the hidden sides are prime numbers.

What is the sum of the hidden prime numbers ?

(A) 15

(B) 20

(C) 25

(D) 30

(E) 35

51.

Which of the following numbers has the smallest prime factor ?

(A) 303

(B) 306

(C) 309

(D) 315

(E) 327

52.

Hannah's mother bought some apples costing more than 10 dollars each at the fruit stand and paid $195. Her farther bought some of the same apples at the same fruit stand and paid $165.

How many apples did mother and farther buy ?

(A) 24

(B) 26

(C) 28

(D) 30

(E) 32

53.

Which of the following numbers is a cube number ?

(A) 2^{2017}

(B) 3^{2018}

(C) 4^{2019}

(D) 5^{2020}

(E) 6^{2021}

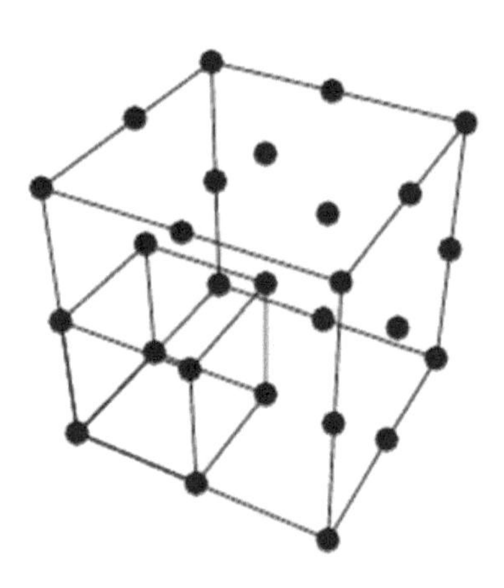

54.

What is the sum of the distinct prime integer divisors of 2025 ?

$(\boldsymbol{A})$ 8

$(\boldsymbol{B})$ 15

$(\boldsymbol{C})$ 22

$(\boldsymbol{D})$ 81

$(\boldsymbol{E})$ 2025

55.

What is the largest power of 3 that is a divisor of $18^4 - 15^4$?

$(\boldsymbol{A})$ 3

$(\boldsymbol{B})$ 9

$(\boldsymbol{C})$ 27

$(\boldsymbol{D})$ 81

$(\boldsymbol{E})$ 243

56.

Suppose that $2^a \times 3^b \times 5^c \times 7^d = 1960$ and a, b, c, and d are positive integer.

What is the value of $2a - 3b + 5c - 7d$?

$(\boldsymbol{A})$ 0

$(\boldsymbol{B})$ -1

$(\boldsymbol{C})$ -2

$(\boldsymbol{D})$ -3

$(\boldsymbol{E})$ Could not be determined.

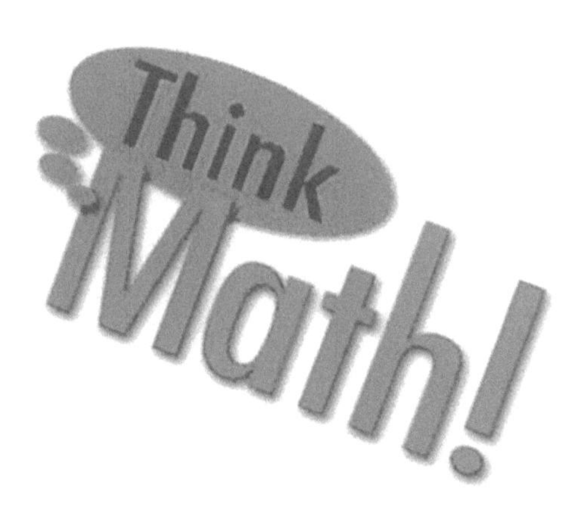

57.

What is the sum of the prime factors of 2020 ?

(A) 100

(B) 102

(C) 104

(D) 106

(E) 108

58.

The Fibonacci Middle School bookstore sells mechanical pencils costing a positive integers of dollars. Some of the 120 eighth graders each bought a mechanical pencil, paying a total of 403 dollars. Some of the 95 seventh graders each bought a mechanical pencil, and they paid a total of 377 dollars.

How many more eighth graders than seventh graders bought a mechanical pencil ?

$(\boldsymbol{A})$ 1

$(\boldsymbol{B})$ 2

$(\boldsymbol{C})$ 3

$(\boldsymbol{D})$ 4

$(\boldsymbol{E})$ 5

59.

The whole numbers A and B are the two smallest whole numbers for which the product of 240 and A is a square and the product of 240 and B is a cube.

What is the quotient of B and A ?

$(\boldsymbol{A})$ 45

$(\boldsymbol{B})$ 60

$(\boldsymbol{C})$ 80

$(\boldsymbol{D})$ 90

$(\boldsymbol{E})$ 100

60.

What is the sum of the two largest prime factors of 816 ?

(A) 14

(B) 16

(C) 18

(D) 20

(E) 22

61.

If $2^a + 128 = 144$, $5^b - 38 = 587$, and, $3^5 + 4^c = 1267$ what is the sum of $a, b,$ and c ?

(A) 9

(B) 11

(C) 13

(D) 15

(E) 17

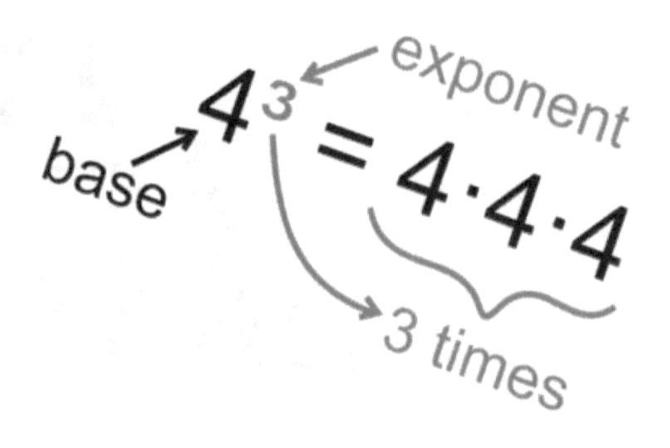

62.

What is the correct ordering of the three numbers, 2^{36}, 5^{18}, 10^{12} ?

$(\boldsymbol{A})$ $10^{12} < 2^{36} < 5^{18}$

$(\boldsymbol{B})$ $5^{18} < 10^{12} < 2^{36}$

$(\boldsymbol{C})$ $5^{18} < 2^{36} < 10^{12}$

$(\boldsymbol{D})$ $2^{36} < 10^{12} < 5^{18}$

$(\boldsymbol{E})$ $2^{36} < 5^{18} < 10^{12}$

PRIMUS INTER PARES
K·DEAN

05. Number of Ways

A. Factorial

For any positive integer n, the product of the integers from 1 to n is called n factorial and is written $n!$;

$$n! = n \times (n-1) \times (n-2) \times \cdots \times 2 \times 1$$

The value of $0!$ is defined to be 1.

$$1! = 1$$
$$2! = 2(1) = 2$$
$$3! = 3(2)(1) = 6$$
$$4! = 4(3)(2)(1) = 24$$
$$5! = 5(4)(3)(2)(1) = 120$$

B. Counting Principle

If one event can occur in a ways, and for each of these ways a second event can occur in b ways, then the number of ways that the two events can occur together is $a \times b$.

The counting principle can be extended to three or more events.

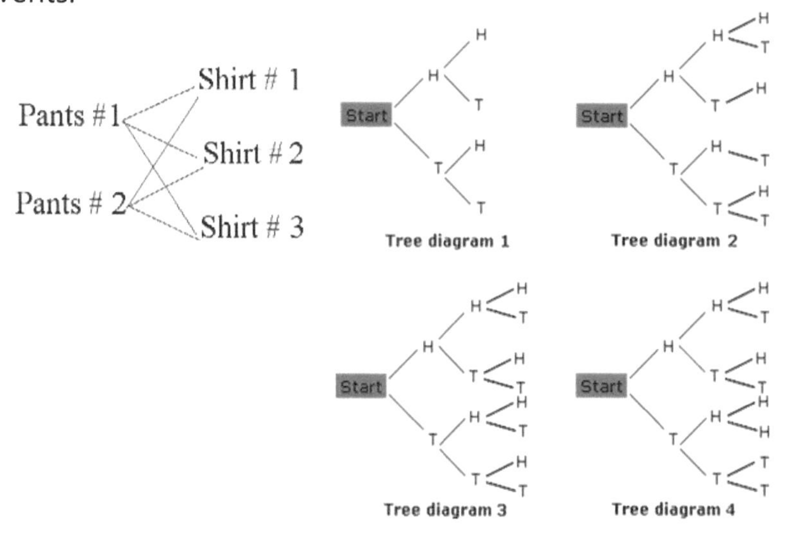

63.

Everyday at school, Lawrence goes up 8 stairs. Lawrence can take the stairs 1, 2, or 3 at a time. For example, Lawrence could go up 3, then 3, then 2.

In how many ways can Lawrence go up the stairs ?

(A) 60

(B) 64

(C) 72

(D) 81

(E) 100

64.

Three Xs, three Ys, and three Zs are placed in the nine spaces so that each row and column contain one of each letter.

		X

If a X is placed in the upper right corner, what is the number of possible arrangements ?

(A) 4

(B) 8

(C) 16

(D) 24

(E) 32

65.

Ten points are spaced around at intervals of one unit around a 2×3 rectangle, as shown as right. Two of the ten points are chosen at random.

What is the probability that the two points are one unit apart?

(A) $\frac{1}{10}$

(B) $\frac{2}{9}$

(C) $\frac{3}{8}$

(D) $\frac{4}{7}$

(E) $\frac{1}{2}$

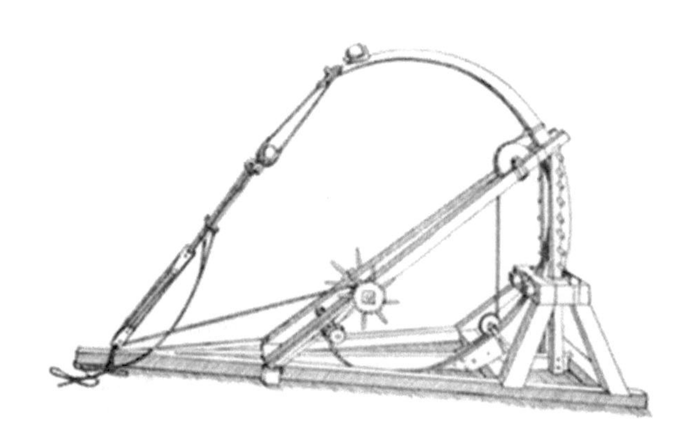

66.

A six-legged Paul has a drawer full of socks, each of which is black, yellow or green, and there are at least six socks of each color. The Paul pulls out one sock at a time without looking.

How many socks must the Paul remove from the drawer to be certain there will be six socks of the same color?

(A) 12

(B) 13

(C) 14

(D) 15

(E) 16

67.

In the arrangement of letters right, by how many different paths can one spell $MATH$?
Beginning at the M in the middle, a path allows only moves from one letter to an adjacent (above, below, left, or right, but not diagonal) letter.
One example of such a path is traced in the picture.

	H	T	H	
H	T	A	T	H
T	A	M	A	T
H	T	A	T	H
	H	T	H	

(A) 15

(B) 18

(C) 20

(D) 24

(E) 30

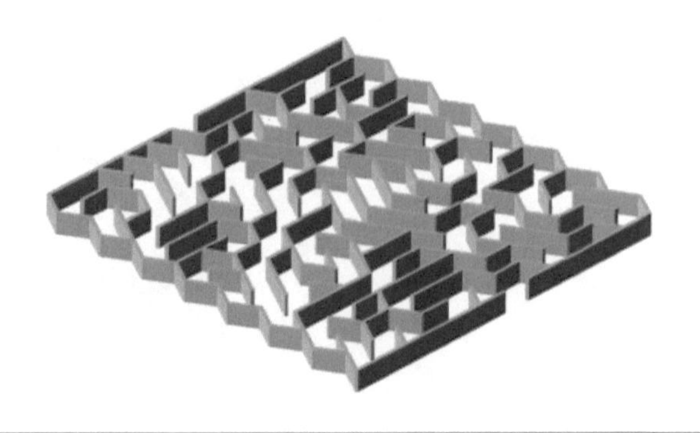

68.

In Olympic qualifying swim competitions, 512 swimmers enter a 50-meter dash competition. The indoor swimming pool has 8 lanes, so only 8 swimmers can compete at a time. At the end of each race, the seven non-winners are eliminated, and the winner will compete again in a later race.

How many races are needed to determine the champion swimmer ?

(A) 10

(B) 15

(C) 64

(D) 68

(E) 73

69.

Becky wants to jog from her house to Peaceful Lake, which is located four blocks west and three blocks south of her house. After jogging each block, Becky can continue either west or south, but she needs to avoid a dangerous construction site two blocks west and two blocks south of her house.

In how many ways can he reach Peaceful Lake by jogging a total of seven blocks?

(A) 12

(B) 16

(C) 17

(D) 18

(E) 20

70.

Douglas lives 3 blocks west and 2 block north of the northwest corner of Central Square. His workplace is 3 blocks east and 3 blocks south of the southeast corner of Central Square. On weekdays he bikes on roads to the northwest corner of Central Square, then takes a diagonal path through the park to the southeast corner, and then bikes on roads to his workplace.

If his route is as short as possible, how many different routes can he take ?

(A) 10

(B) 30

(C) 200

(D) 300

(E) 300

71.

A three-digit integer contains one of each of the digits 0, 2, and 4.

What is the probability that the integer is divisible by 10 ?

$(A)\ \frac{1}{6}$

$(B)\ \frac{1}{4}$

$(C)\ \frac{1}{3}$

$(D)\ \frac{1}{2}$

$(E)\ \frac{2}{3}$

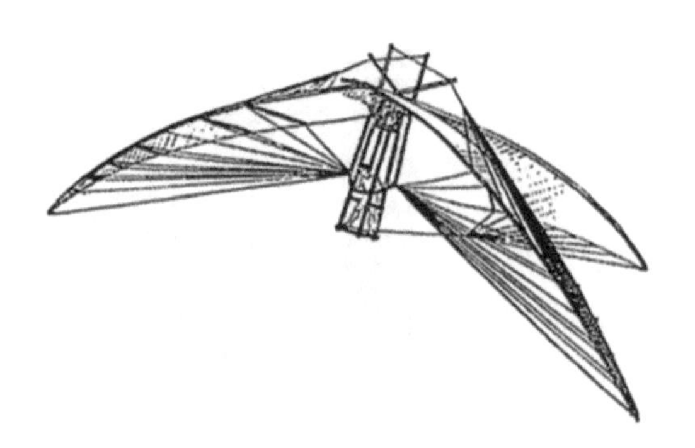

72.

At the FIFA World Cup, the losing team of each game is eliminated from the tournament.

If thirty two teams compete, how many games will be played to determine the winner ?

(A) 24

(B) 25

(C) 30

(D) 31

(E) 32

06. Average

Definition

The mean is the average of the numbers. It is easy to calculate: add up all the numbers, then divide by how many numbers there are. In other words it is the sum divided by the count.

$$\text{average} = \frac{\text{sum of values}}{\text{number of values}}.$$

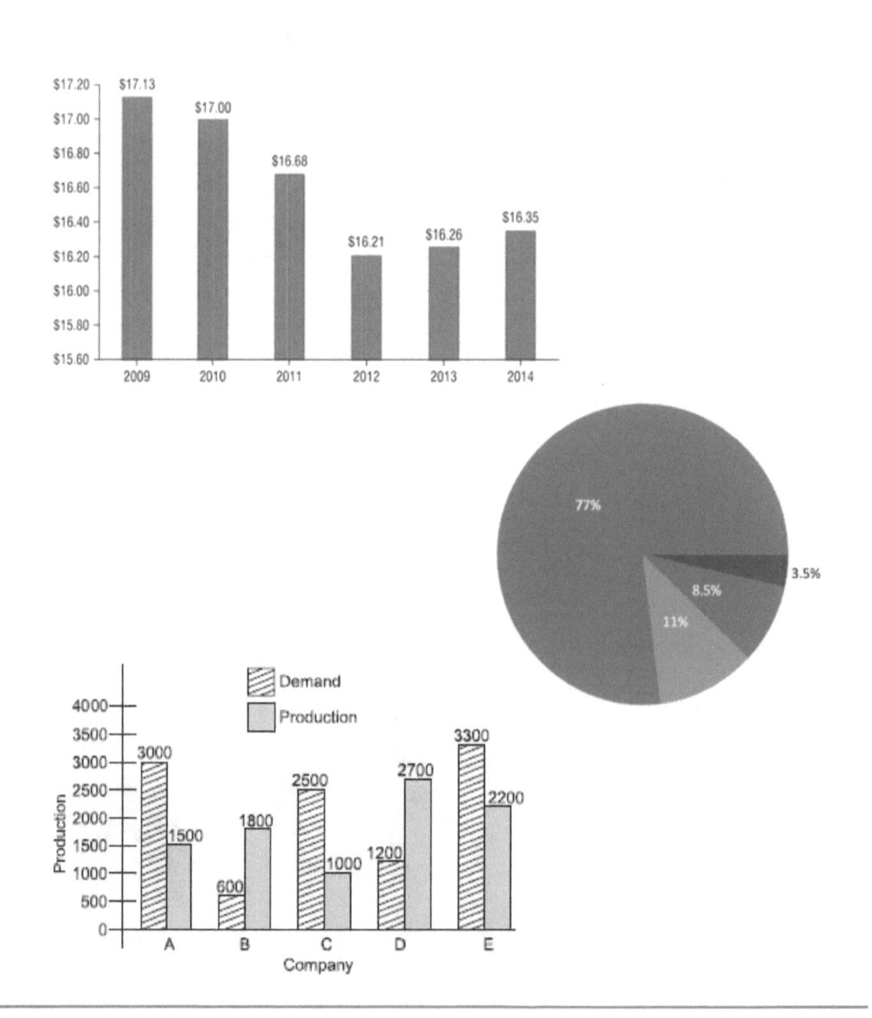

73.

Peter and Rebecca have a same rectangular array of integers with 65 rows and 30 columns. Peter adds the integers in each row and the calculates the total sum of 65 rows. Rebecca adds the numbers in each column and calculates the total sum of 30 columns.

What is the ratio of the average of row's sum to the average of column's sum ?

(A) 1

(B) $\frac{13}{6}$

(C) $\frac{6}{13}$

(D) $\frac{7}{12}$

(E) $\frac{12}{7}$

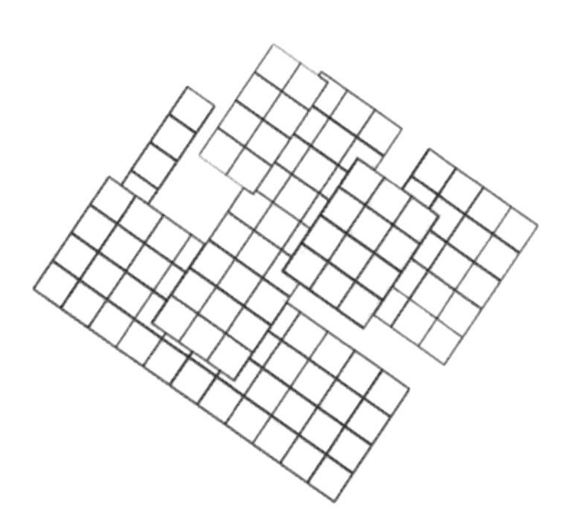

74.

In Donald's first seven handball games, he scored $5, 8, 2, 4,$ $7, 5,$ and 4 points. In his eighth game, he scored fewer than 9 points and his points-per-game average for the eight games was an integer and in his ninth game, he scored fewer than 9 points and his points-per-game average for the nine games was also an integer . Similarly in his tenth game, he scored fewer than 9 points and his points-per-game average for the ten games was also an integer.

What is the sum of the number of points she scored in the eighth, ninth and tenth games?

$(\boldsymbol{A})$ 11

$(\boldsymbol{B})$ 12

$(\boldsymbol{C})$ 13

$(\boldsymbol{D})$ 14

$(\boldsymbol{E})$ 15

75.

Five students take a geometry exam. Four of their scores are $65, 75, 80$, and 90.

If the average of their five scores is 77, then what is the fifth student's score?

(A) 75

(B) 76

(C) 77

(D) 78

(E) 79

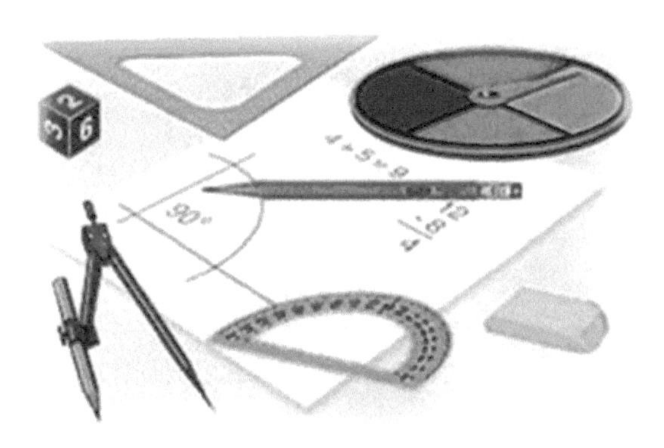

76.

Edwin must take five 100-point tests in his chemistry class. His goal is to achieve an average grade of 90 on the tests. His first three test scores were 84, 93, and 90. After seeing his score on the fourth test, he realized he can still reach his goal.

What is the lowest possible score he could have made on the fourth test ?

(A) 83

(B) 85

(C) 87

(D) 89

(E) 91

77.

The double bar graph shows the number of minutes of using smart device by both Bessie (black bar) and Chloe (white bar) in one week.

On the mean, how many more minutes per day did Bessie use the smart device than Chloe ?

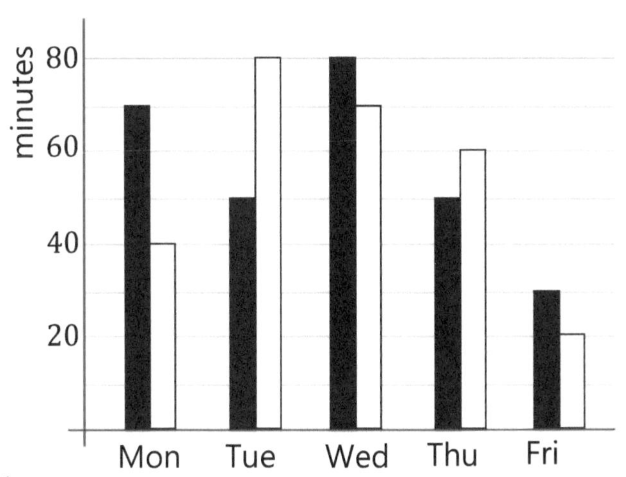

(A) 1

(B) 2

(C) 3

(D) 4

(E) 5

78.

Orange sales from the Low Hill Farm from Monday through Friday are shown .

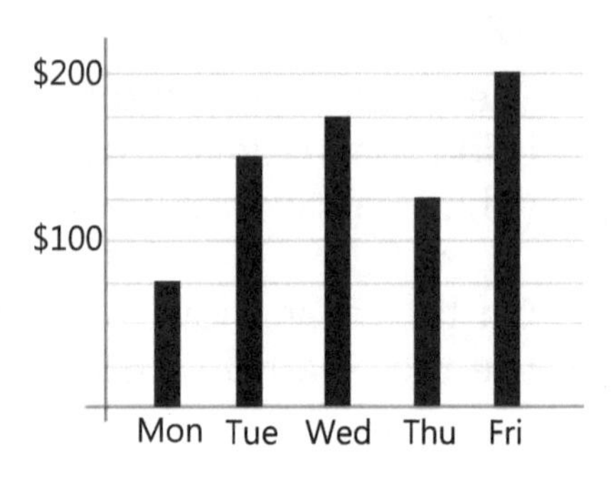

What were the average sales per day in dollars ?

(A) 135

(B) 140

(C) 145

(D) 150

(E) 155

79.

The mean age of the 7 people in Room No. 1 is 36. The mean age of the 5 people in Room No. 2 is 42.

If they meet all together at Room No. 3, what is the mean age of all the people ?

(A) 38.0

(B) 38.5

(C) 39.0

(D) 39.5

(E) 40.0

80.

Rachel's parents have agreed to buy her tickets to see BTS band if she spends an average of 14 hours per week studying mathematics for 5 weeks. For the first four weeks she studies mathematics for 13, 15, 11, and 14 hours.

How many hours must she study for the final week to earn the tickets?

(A) 14

(B) 15

(C) 16

(D) 17

(E) 18

81.

The average age of 7 people in a room is 35 years. An 11-year-old boy leaves the room.

What is the average age of the six remaining people ?

$(\boldsymbol{A})$ 37

$(\boldsymbol{B})$ 38

$(\boldsymbol{C})$ 39

$(\boldsymbol{D})$ 40

$(\boldsymbol{E})$ 41

82.

The average of the six numbers in a list is 42. The average of the first three numbers is 36.

What is the average of the last three numbers?

(A) 44

(B) 46

(C) 48

(D) 50

(E) 52

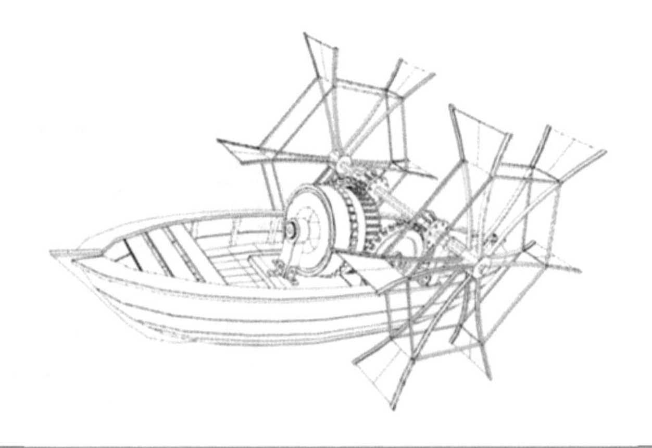

83.

Maggie and Owen each took five 100-point math tests. Maggie averaged 89 on the five tests. Owen scored 8 points lower than Maggie on the first test, 12 points higher than her on the second test, 11 points higher than her on the third test and 15 points lower on both the fourth and fifth tests.

What is the difference between Maggie's average and Owen's average on these five tests?

(A) 3

(B) 4

(C) 5

(D) 6

(E) 7

84.

What is the smallest possible average of six distinct positive odd integers?

(A) 4

(B) 5

(C) 6

(D) 7

(E) 8

85.

Jason jogged 1 hour 20 min each day for 5 days and 1 hour 40 min each day for 4 days.

How long would he have to jog the tenth day in order to average 90 minutes of jogging each day for the entire time ?

(A) 1 hour 20 minutes

(B) 1 hour 25 minutes

(C) 1 hour 30 minutes

(D) 1 hour 35 minutes

(E) 1 hour 40 minutes

PRIMUS INTER PARES
K·DEAN

Answer and Explanations

01. Answer (*D*)

Considering the multiples of each number.

Let there be t tomatoes, p pineapples and w watermelons;

$$t+p+w=15 \quad and \quad t+4p+5w=25 \quad and \quad t,p,w \geq 1$$

Finding the difference of the two equations;

$$3p+4w=10$$

Because p and w are whole numbers greater than or equal to 1,

if $w=3$, then $4w=12>10$ → impossible,

if $w=2$, then $4w=8 \rightarrow 3p+8=10 \rightarrow 3p=2$

$\rightarrow p=\frac{2}{3}$(not whole number) → impossible.

if $w=1$, then $4w=4 \rightarrow 3p+4=10 \rightarrow 3p=6$

$\rightarrow p=2$ → it is possible.

So, $w=1$, $p=2$ and $t=12$ ($\because t+p+w=15$)

Thus, the correct answer is ***D***.

02. Answer (*B*)

Adding the weight of pairs.

Let A, B and C be the weight of each box so, the weight of pairs are;

$$A+B=7.5, \qquad A+C=8.5, \qquad B+C=8$$

Adding all pairs are;

$$A+B+A+C+B+C=7.5+8.5+8=24$$

$$\rightarrow 2(A+B+C)=24 \rightarrow A+B+C=12$$

Thus, the correct answer is ***B***.

03. Answer (*C*)

Setting up a linear equation.

Let w be the number of weeks;

$$700=80+70+100+50w \rightarrow 50w+250=700$$

$$\rightarrow 50w=450 \rightarrow w=9$$

Thus, the correct answer is ***C***.

04. **Answer** $(\boldsymbol{B})$

Setting up a linear equation.

Let x be the total amount of oranges;

$$x = \frac{1}{3}x + 5 + 3 \quad \rightarrow \quad \frac{2}{3}x = 8 \quad \rightarrow \quad x = 8 \times \frac{3}{2} = 12$$

Thus, the correct answer is $\boldsymbol{B}$.

05. **Answer** $(\boldsymbol{C})$

The number of winning games is equal to losing.

The number of winning games is $3 + 2 + 2 = 7$ games and $1 + 4 + x = 7$ games. So, the number of her losing game is 2.

Thus, the correct answer is $\boldsymbol{C}$.

06. **Answer** $(\boldsymbol{D})$

Make a linear equation.

Let a be the amount of apples and b the number of baskets, then

$$a = 12b - 12 \times 3 = 8b + 4 \quad \rightarrow \quad 12b - 36 = 8b + 4$$
$$\rightarrow \quad 4b = 40 \quad \rightarrow \quad b = 10 \ \ and \ a = 8 \times 10 + 4 = 84$$

Thus, the correct answer is $\boldsymbol{D}$.

07. **Answer** $(\boldsymbol{D})$

It is a linear equation.

From the given definition,

$$3 * a = 3 - 2a$$

So,

$$a * (3 * a) = a * (3 - 2a) = a - 2(3 - 2a) = a - 6 + 4a = 5a - 6$$

Therefore,

$$a * (3 * a) = 4 \quad \rightarrow \quad 5a - 6 = 4 \quad \rightarrow \quad 5a = 10 \quad \rightarrow \quad a = 2$$

Thus, the correct answer is $\boldsymbol{D}$.

08. **Answer** $(\boldsymbol{D})$

Let x be the number of pages of the novel.

Let x be the number of pages of the novel;

① the remain pages of first week;

$$x-\left(\frac{1}{6}x+25\right)=\frac{5}{6}x-25$$

② the remain pages of second week;

$$\left(\frac{5}{6}x-25\right)-\left(\frac{1}{5}\left(\frac{5}{6}x-25\right)+32\right)=\frac{5}{6}x-25-\frac{1}{6}x+5-32$$
$$=\frac{4}{6}x-52$$

③ the remain pages of third week;

$$\left(\frac{4}{6}x-52\right)-\left(\frac{1}{4}\left(\frac{4}{6}x-52\right)+15\right)=\frac{4}{6}x-52-\frac{1}{6}x+13-15$$
$$=\frac{3}{6}x-54=\frac{1}{2}x-54$$

④ 78 pages left to read, which he read the fourth week.

$$\rightarrow \frac{1}{2}x-54=78 \quad\rightarrow\quad \frac{1}{2}x=132 \quad\rightarrow\quad 264$$

So, the number of pages in the novel is 264.

Thus, the correct answer is $\boldsymbol{D}$.

09. **Answer** $(\boldsymbol{A})$

Set up a linear system.

Let x be the number of birds, y be the number of mammals;

① counted 124 heads

$$x+y=124 \quad\rightarrow\quad ①$$

② counted ~326 legs

$$2x+4y=326 \quad\rightarrow\quad ②$$

Solving the linear system by ② − ① × 2

$$2x+4y=326$$
$$2x+2y=248$$
$$\rightarrow\quad 2y=78 \quad\rightarrow\quad y=39$$

Thus, the correct answer is $\boldsymbol{A}$.

10. Answer (A)

Setting up a linear system.

Let x be the number of bicycles and y be the number of tricycle;
① counted 8 students → $x + y = 8$
② counted ~ 21 wheels → $2x + 3y = 21$
So, $3 \times$ ① − ② vertically;

$$3x + 3y = 24$$
$$2x + 3y = 21$$

Therefore $x = 3$

Thus, the correct answer is $\boldsymbol{A}$.

11. Answer (B)

Set up a linear system.

Let x be the number of correct answers and y be the number of incorrect answers and set up a linear system;
① quiz with fifteen problems; $x + y = 15$
② three points for a correct answer and ~ one point for an incorrect answer. ~ his score was 25; $3x - y = 25$
By ①× 3 −② vertically;

$$3x + 3y = 45$$
$$3x - \ y = 25$$
$$4y = 20 \ \rightarrow \ y = 5$$

Thus, the correct answer is $\boldsymbol{B}$.

12. Answer (D)

Considering the multiples of each number.

Let there be t tomatoes, p pineapples and w watermelons;

$$t + p + w = 15 \quad and \quad t + 4p + 5w = 25 \quad and \quad t, p, w \geq 1$$

Finding the difference of the two equations;

$$3p + 4w = 10$$

Because p and w are whole numbers greater than or equal to 1,
if $w = 3$, then $4w = 12 > 10$ → impossible,
if $w = 2$, then $4w = 8$ → $3p + 8 = 10$ → $3p = 2$
→ $p = \frac{2}{3}$(not whole number) → impossible.
if $w = 1$, then $4w = 4$ → $3p + 4 = 10$ → $3p = 6$
→ $p = 2$ → it is possible.
So, $w = 1$, $p = 2$ and $t = 12$ ($\because t + p + w = 15$)

Thus, the correct answer is $\boldsymbol{D}$.

13. **Answer** $(\boldsymbol{E})$

Finding price for a ticket.

Each of his five friend friends paid \$1.4 to cover his portion. So the price for a movie ticket is $5 \times \$1.40 = \$7$.
Therefore, the movie total price is

$$6 \times \$7 = \$42.00$$

Thus, the correct answer is $\boldsymbol{E}$.

14. **Answer** $(\boldsymbol{B})$

Using pennies or quarters.

① largest number of coins; using only pennies

$$\$0.89 = 89\ cents = 89\ pennies \quad \rightarrow \quad 89\ coins$$

② smallest number of coins; using 3 quarters

$$\begin{aligned} 89\ cents = &\ 3\ quarters(3 \times 25 cents \rightarrow 75 cents) \\ &+ 1\ dime\ (1 \times 10\ cents \rightarrow 10 cents) \\ &+ 4\ pennies(4 \times 1 cent \rightarrow 4\ cents) \\ &\rightarrow 8\ coins \end{aligned}$$

From ① and ②, the sum is $89 + 8 = 97$ coins.

Thus, the correct answer is $\boldsymbol{B}$.

15. **Answer** $(\boldsymbol{C})$

Set up the two variable for the number of students.

Let x be 6th grade students and y be 7th grade students;
'the same number of 6th grade students and 7th grade students'

$$\rightarrow \quad x = y$$

And, 'fraction of the students were 6th grade' is

$$\frac{6th\ grade}{total\ student} = \frac{\frac{4}{5}x}{\frac{4}{5}x + \frac{3}{4}y} = \frac{\frac{4}{5}x}{\frac{4}{5}x + \frac{3}{4}x} = \frac{\frac{4}{5}x}{\frac{31}{20}x} = \frac{16}{31}$$

Thus, the correct answer is $\boldsymbol{C}$.

16. Answer (*B*)

Set up a quadratic equation.

Let x be the number of boys so, the number of girls is $x-1$. He gave each boy x lollipops and each girl $x-1$ lollipops so the total lollipops are $x^2+(x-1)^2$.

Therefore,

$$x^2+(x-1)^2=50-9 \rightarrow x^2+x^2-2x+1=41$$
$$\rightarrow 2x^2-2x-40=0$$
$$\rightarrow x^2-x-20=0$$

By factorization,

$$(x-5)(x+4)=0 \rightarrow x=5 \ or \ x=-4$$

x is positive so, the number of boys is 5 and the number of girls is 4.

Thus, the correct answer is $\boldsymbol{B}$.

17. Answer (*A*)

Drawing a Venn diagram.

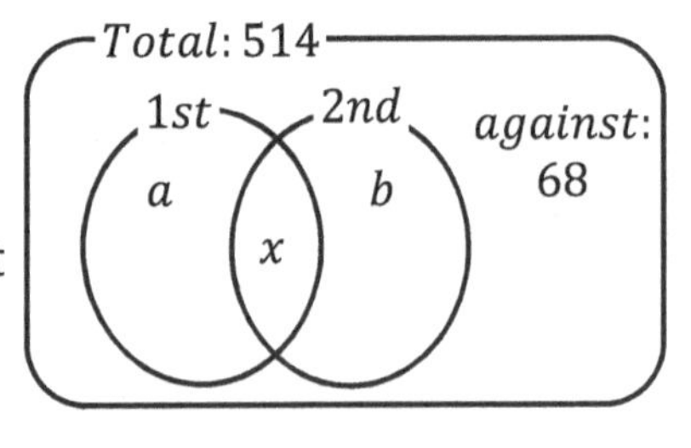

From the Venn diagram;

$a+b+x=514-68=446$

And 325 voted in favor of the first issue so, $a+x=325$.

$a+x+b=446 \rightarrow 325+b=446$

$\rightarrow b=121$

From 272 voted in favor of the second issue,

$$b+x=272 \rightarrow 121+x=272 \rightarrow x=151$$

So, x, the number of people voted in favor of both issues, is 151.

Thus, the correct answer is $\boldsymbol{A}$.

18. Answer (*C*)

Drawing a Venn diagram.

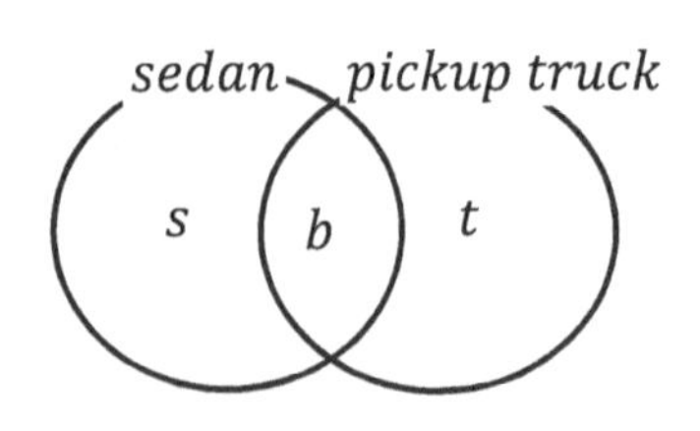

① 405 families

$\rightarrow s+b+t=405$

② 218 families own pickup trucks

$\rightarrow b+t=218$

$\rightarrow s+218=405 \rightarrow s=187$

Thus, the correct answer is $\boldsymbol{C}$.

19. Answer (*E*)

Draw the Venn diagram.

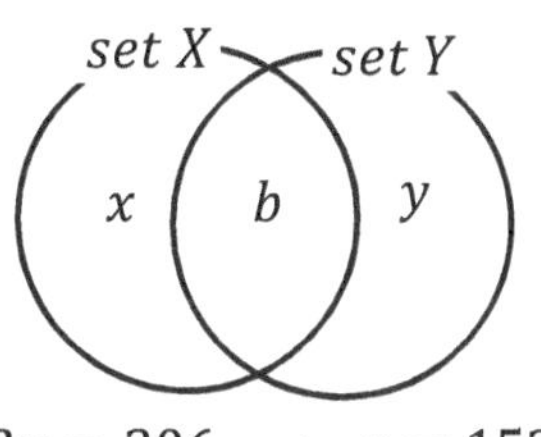

Draw the Venn diagram.
union ~ has 818: $x + b + y = 818$
intersection ~ 512: $b = 512$

So,

$x + 512 + y = 818 \rightarrow x + y = 306 \rightarrow 2y = 306 \rightarrow y = 153$

Therefore, $b + y = 512 + 153 = 665$

Thus, the correct answer is ***E***.

20. Answer (*C*)

Zero is a whole number.

Let x and y be the added two numbers, then;

$$\frac{55 + x + y}{13} = 5 \rightarrow 55 + x + y = 65 \rightarrow x + y = 10.$$

So, the sum of two elements is 10 and the subsets are;

$$\{0,10\}, \{1,9\}, \{2,8\}, \{3,7\}, \{4,6\}$$

Thus, the correct answer is ***C***.

21. Answer (*E*)

Drawing the given situation.

From the figure, the distance is the length of the diagonal, so using the Pythagorean theorem;

$$d = \sqrt{\left(\frac{2}{3}\right)^2 + \left(\frac{2}{3} + \frac{1}{3}\right)^2} = \sqrt{\frac{13}{9}} = \frac{\sqrt{13}}{3}$$

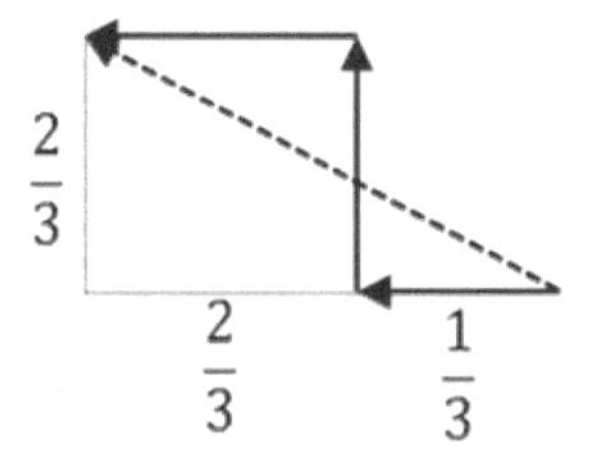

Thus, the correct answer is ***E***.

22. Answer (*A*)

Using the Pythagorean theorem.

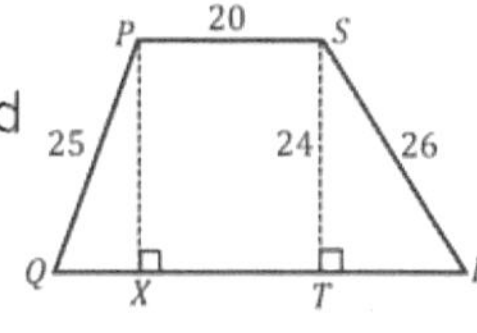

Drawing a perpendicular line PX, so $PX = 24$, and using the Pythagorean theorem;

$$RT = \sqrt{26^2 - 24^2} = 10$$

$$QX = \sqrt{25^2 - 24^2} = 7$$

Therefore, the perimeter is $20 + 26 + 10 + 20 + 7 + 25 = 108$.

Thus, the correct answer is ***A***.

23. **Answer** $(\boldsymbol{B})$

Using the Pythagorean theorem.

From the Pythagorean theorem;

$$13^2 = 12^2 + 5^2 \quad \rightarrow \quad C = A + B \quad \rightarrow \quad A = C - B$$

Thus, the correct answer is $\boldsymbol{B}$.

24. **Answer** $(\boldsymbol{C})$

The given triangle is an isosceles triangle.

The given triangle is an isosceles triangle and point X is midpoint of base PR, so using the Pythagorean theorem;

$$OX = \sqrt{OP^2 - PX^2} = \sqrt{13^2 - 5^2} = 12$$

Therefore the area is;

$$area = \frac{1}{2} \times base \times height = \frac{1}{2} \times 10 \times 12 = 60$$

Or, using the Heron's formula;

$$s = \frac{1}{2} \times perimeter = \frac{1}{2} \times (13 + 13 + 10) = 18$$

$$area = \sqrt{s(s-13)(s-13)(s-10)}$$

$$= \sqrt{18(18-13)(18-13)(18-10)} = 60$$

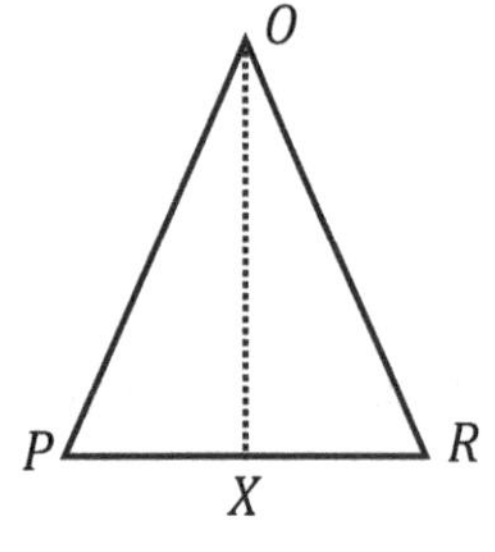

Thus, the correct answer is $\boldsymbol{C}$.

25. **Answer** $(\boldsymbol{D})$

Using the Pythagorean theorem.

① Area of □PQRS; $area = 4 \times 7 = 28$

② From △OPQ,

$$area = \frac{1}{2} \times OP \times PQ = \frac{1}{2} \times OP \times 7 = 28 \quad \rightarrow \quad OP = 8$$

Using the Pythagorean theorem;

$$OQ = \sqrt{7^2 + 8^2} = \sqrt{113}$$

Thus, the correct answer is $\boldsymbol{D}$.

26. **Answer** ($\boldsymbol{D}$)

Using the Pythagorean theorem.

① The side length of AB is; $64 = \overline{AB}^2 \rightarrow \overline{AB} = 8$ and
② The side length of AC is; $68 = 4 \times \overline{AC} \rightarrow \overline{AC} = 17$.
△ABC is a right angular triangle and from ① and ②;

$$\overline{BC} = \sqrt{\overline{AC}^2 - \overline{AB}^2} = \sqrt{17^2 - 8^2} = \sqrt{225} = 15$$

Thus, the correct answer is $\boldsymbol{D}$.

27. **Answer** ($\boldsymbol{D}$)

There are four right triangles.

① The side length of the larger square is $\sqrt{196} = 14$.
② The length of short segment, a, and the length of long segment, b, is

$$a = 14 \times \frac{3}{3+4} = 6, \qquad b = 14 \times \frac{4}{3+4} = 8$$

③ The side length, c, of the smaller square is calculated by the Pythagorean theorem;

$$c = \sqrt{a^2 + b^2} = \sqrt{6^2 + 8^2} = 10$$

So, the area of the smaller square is $10 \times 10 = 100$.

Thus, the correct answer is $\boldsymbol{D}$.

28. **Answer** ($\boldsymbol{B}$)

Using the Pythagorean theorem.

The two triangles are isosceles triangles so, find these height by using the Pythagorean theorem.

① 26-26-20 triangle

$$x = \sqrt{26^2 - 10^2} = 24$$

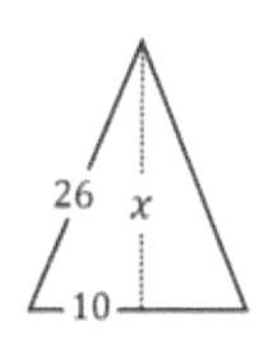

Area is;

$$\frac{1}{2} \times 20 \times x = \frac{1}{2} \times 20 \times 24 = 240$$

② 26-26-48 triangle

$$y = \sqrt{26^2 - 24^2} = 10$$

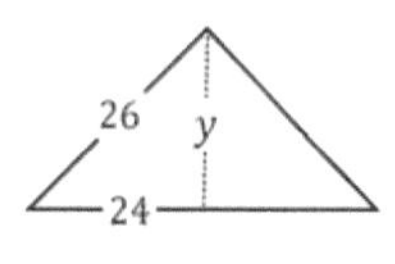

Area is;

$$\frac{1}{2} \times 48 \times y = \frac{1}{2} \times 48 \times 10 = 240$$

So, $Y - X = 240 - 240 = 0$

Thus, the correct answer is $\boldsymbol{B}$.

29. Answer (D)

There are two squares.

Let s be the side length of the extended triangle, then the length of hypotenuse is
$\sqrt{s^2+s^2}=\sqrt{2s^2}=\sqrt{2}s$.
So, the area of
① is; $\frac{1}{2}\times s\times s=\frac{1}{2}s^2$,
② is; $s\times\sqrt{2}s=\sqrt{2}s^2$
③ is $\sqrt{2}s\times\sqrt{2}s=2s^2$
So, $(8\times① + 4\times② + ③) - 8\times① = 4\times\sqrt{2}s^2+2s^2=(2+4\sqrt{2})s^2$.

Thus, the correct answer is $\boldsymbol{D}$.

(A)

Using the Pythagorean theorem.

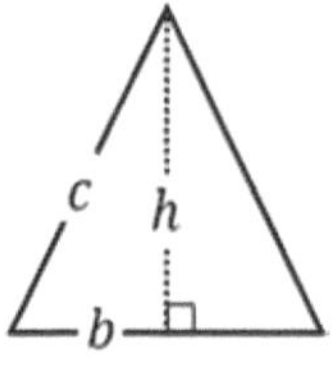

The area is 48, so

$$48=\frac{1}{2}\times(2b)\times h \quad\rightarrow\quad 8h=48 \quad\rightarrow\quad h=6$$

Using the Pythagorean theorem;

$$c=\sqrt{b^2+h^2}=\sqrt{64+36}=10$$

So, the perimeter is $2b+c+c=16+10+10=36$.

Thus, the correct answer is $\boldsymbol{A}$.

31. Answer (B)

Using the Pythagorean theorem.

The side length of PQ and QR are 6 unit, so use the Pythagorean theorem;

$$PR=\sqrt{6^2+6^2}=6\sqrt{2}$$

Therefore, the side length of RX is;

$$\frac{2}{3}\times 6\sqrt{2}=4\sqrt{2}$$

Thus, the correct answer is $\boldsymbol{B}$.

32. **Answer** $(\boldsymbol{B})$

The triangle is a right triangle.

From the property of circle, the triangle is a right triangle and $PR = PQ$, so

$$\frac{1}{2} \times PR \times PQ = 2 \quad \rightarrow \quad PR^2 = 4 \quad \rightarrow \quad PR = 2.$$

Using the Pythagorean theorem;

$$QR = \sqrt{2^2 + 2^2} = 2\sqrt{2}.$$

Using the formula of area of triangle;

$$\frac{1}{2} \times QR \times OP = \frac{1}{2} \times 2\sqrt{2} \times OP = 2 \quad \rightarrow \quad OP = \frac{2}{\sqrt{2}}.$$

Therefore, the area of semicircle is;

$$\frac{1}{2}\pi r^2 = \frac{1}{2} \times \pi \times \left(\frac{2}{\sqrt{2}}\right)^2 = \frac{1}{2} \times \pi \times \frac{4}{2} = \pi$$

Thus, the correct answer is $\boldsymbol{B}$.

33. **Answer** $(\boldsymbol{B})$

The area of three triangles is same.

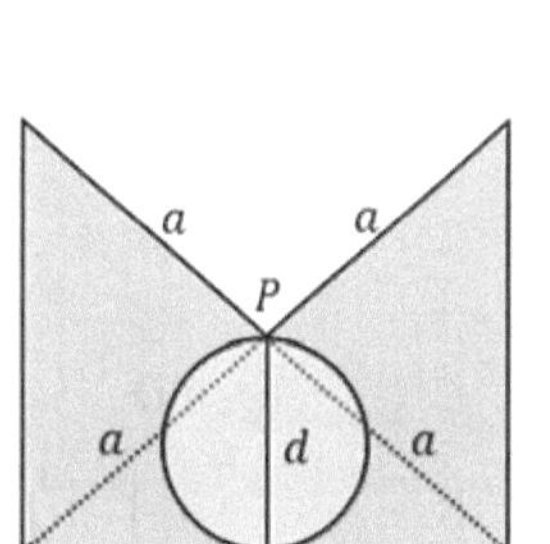

The point P is midpoint of hypotenuse and the angle $\angle P$ is $90°$,
so using the Pythagorean theorem;

$$2a = \sqrt{4^2 + 4^2} = 4\sqrt{2} \quad \rightarrow \quad a = 2\sqrt{2}$$

The area of three triangles is same, so

$$\frac{1}{2} \times a \times a = \frac{1}{2} \times 4 \times d \quad \rightarrow \quad 2d = \frac{1}{2} \times \left(2\sqrt{2}\right)^2$$
$$\rightarrow \quad d = 2$$

Using the area of triangle, the area is

$$3 \times \frac{1}{2} \times a \times a - \pi\left(\frac{d}{2}\right)^2 = 12 - \pi$$

Thus, the correct answer is $\boldsymbol{B}$.

34. **Answer** (E)

The given triangle is a right triangle.

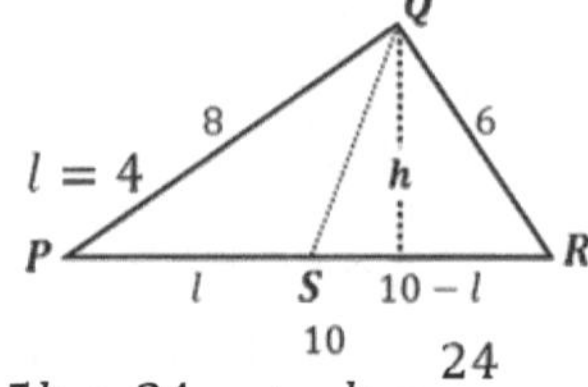

Let l be the length of $\overline{PS}$, then
$QS + 8 + l = QS + 6 + 10 - l \ \rightarrow \ 2l = 8 \ \rightarrow \ l = 4$
Since 6, 8, 10 is Pythagorean triple, and let h be the height of $\triangle PQR$;

$$area\ of\ PQR = \frac{1}{2} \times 6 \times 8 = \frac{1}{2} \times 10 \times h \ \rightarrow \ 5h = 24 \ \rightarrow \ h = \frac{24}{5}$$

So, the area of $\triangle PQS$ is;

$$area = \frac{1}{2} \times 4 \times \frac{24}{5} = \frac{48}{5} = 9.6$$

Thus, the correct answer is $\boldsymbol{E}$.

35. **Answer** (C)

Using the Pythagorean theorem.

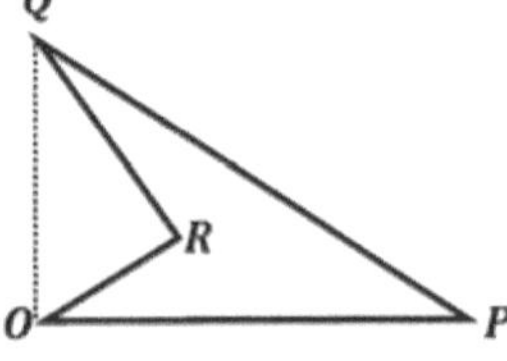

Drawing the line $\overline{OQ}$, then $\triangle OQR$ is a right triangle and $\overline{OQ} = 10(\because 10^2 = 6^2 + 8^2)$.
Since $26^2 = 24^2 + 10^2$, $\triangle OPQ$ is also a right triangle, so the area of $\square OPQR$ is;

$$\triangle OPQ - \triangle OQR = \tfrac{1}{2} \times 24 \times 10 - \tfrac{1}{2} \times 6 \times 8 = 96.$$

Thus, the correct answer is $\boldsymbol{C}$.

36. **Answer** (D)

Using the Pythagorean theorem.

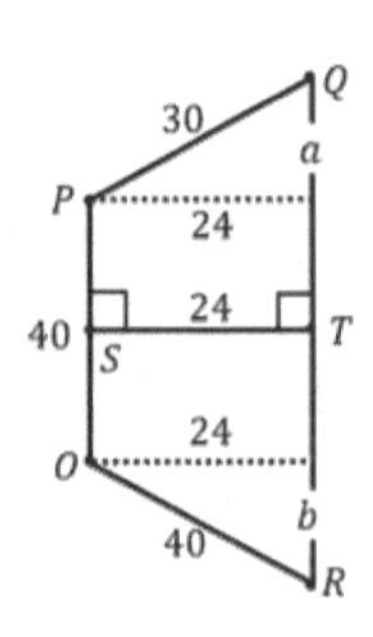

① Finding a and b by Pythagorean theorem;

$$a = \sqrt{30^2 - 24^2} = \sqrt{324} = 18$$

$$b = \sqrt{40^2 - 24^2} = \sqrt{1024} = 32$$

② The area of the trapezoid is;

$$\frac{1}{2} \times (40 + 40 + 18 + 32) \times 24 = 1560$$

Thus, the correct answer is $\boldsymbol{D}$.

37. Answer $(\boldsymbol{B})$

Using the Pythagorean theorem.

Using the Pythagorean theorem;

$$CE = \sqrt{5^2 - 4^2} = 3$$

So, the area is;

$$\frac{1}{2} \times (4 + 7) \times 4 = 22$$

Thus, the correct answer is $\boldsymbol{B}$.

38. Answer $(\boldsymbol{C})$

Using the relation of the area.

The area of $\square XYWZ$ is calculated by;

$$10 \times 10 - \left(2 \times \frac{1}{2} \times 8 \times 6 + 2 \times \frac{1}{2} \times 4 \times 2\right) = 44.$$

And using the Pythagorean theorem, the length of $\overline{YW}$ is;

$$YW = \sqrt{8^2 + 6^2} = 10.$$

Because d is height of $\square XYWZ$, the area is;

$$area = d \times \overline{YW} \quad \rightarrow \quad d \times 10 = 44 \quad \rightarrow \quad d = 4.4$$

Thus, the correct answer is $\boldsymbol{C}$.

39. Answer $(\boldsymbol{C})$

Finding one side length of the square.

Let x be the one side length of the square, the value of x is calculated by using the Pythagorean theorem;

$$\sqrt{x^2 + x^2} = 4^2 \quad \rightarrow \quad x^2 = 8 \quad \rightarrow \quad x = 2\sqrt{2}$$

And the areas of all three squares is 8.

A figure

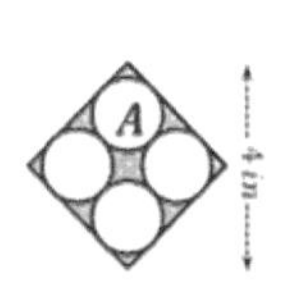

- the radius of the small circle is $\frac{2\sqrt{2}}{4} = \frac{\sqrt{2}}{2}$ and

  the area of 4 circles is $4 \times \pi r^2 = 4 \times \pi \times \left(\frac{\sqrt{2}}{2}\right)^2 = 2\pi$

  So, the area of shaded region is $8 - 2\pi$ - ①

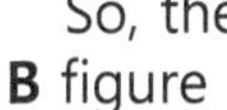

B figure

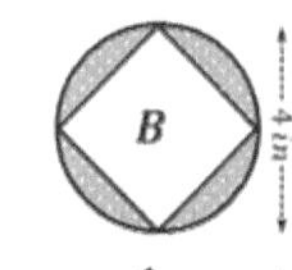

- the radius of the circle is $\frac{4}{2} = 2$ and the area of circle is $\pi r^2 = \pi \times (2)^2 = 4\pi$

  So, the area of shaded region is $4\pi - 8$ - ②

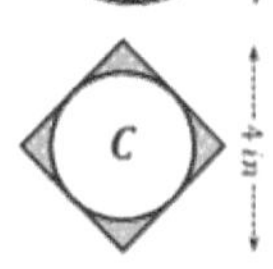

C figure

- the radius of circle is $\frac{2\sqrt{2}}{2} = \sqrt{2}$ and

  the area of circle is $\pi r^2 = \pi \times \left(\sqrt{2}\right)^2 = 2\pi$

  So, the area of shaded region is $8 - 2\pi$ - ③

From ①, ②, and ③, the sum is $8 - 2\pi + 4\pi - 8 + 8 - 2\pi = 8$

Thus, the correct answer is $\boldsymbol{C}$.

40. **Answer** **(*C*)**

Using the Pythagorean theorem.

The midpoint of width lies on the center of the circle, so the radius is calculated by using the Pythagorean theorem.

$$r = \sqrt{3^2 + 1^2} = \sqrt{10}$$

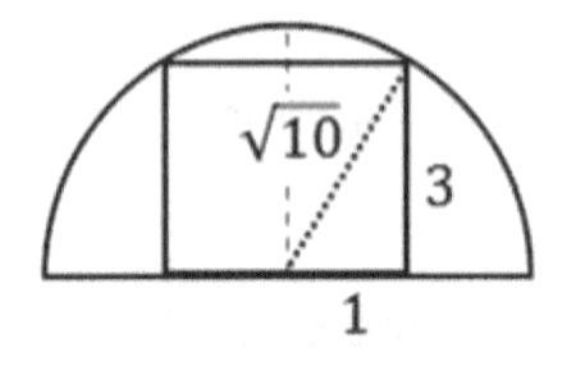

Therefore the area of the semicircle is;

$$area = \frac{1}{2} \times \pi r^2 = \frac{1}{2} \times \pi \times \left(\sqrt{10}\right)^2 = 5\pi$$

Thus, the correct answer is ***C***.

41. **Answer** **(*E*)**

Using the Pythagorean theorem.

By the Pythagorean theorem;

$$R^2 = r^2 + r^2 \rightarrow 2r^2 = 4 \rightarrow r^2 = 2 \rightarrow r = \sqrt{2}$$

So,

① the area of shaded region is;

$$(2r)^2 - \pi r^2 = \left(2\sqrt{2}\right)^2 - \pi\left(\sqrt{2}\right)^2 = 8 - 2\pi$$

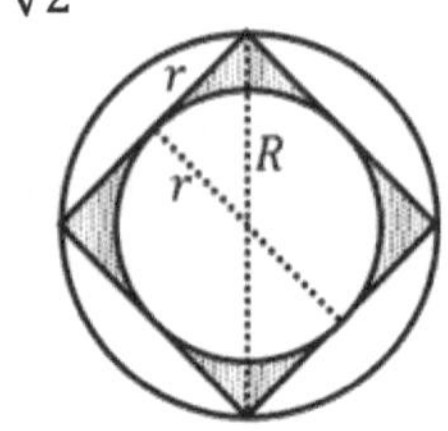

② the area between the two circles is;

$$\pi R^2 - \pi r^2 = \pi(2)^2 - \pi\left(\sqrt{2}\right)^2 = 2\pi$$

From ① and ②, the fraction is;

$$\frac{8 - 2\pi}{2\pi} = \frac{4 - \pi}{\pi}$$

Thus, the correct answer is ***E***.

42. **Answer** **(*D*)**

The shaded region is the area between the two circles.

Point B is mid point of AB and using the Pythagorean theorem;

$$OA = \sqrt{AB^2 + OB^2} = \sqrt{8^2 + 6^2} = \sqrt{100} = 10$$

So, the area of the shaded region is;

$$100\pi - 36\pi = 64\pi$$

Thus, the correct answer is ***D***.

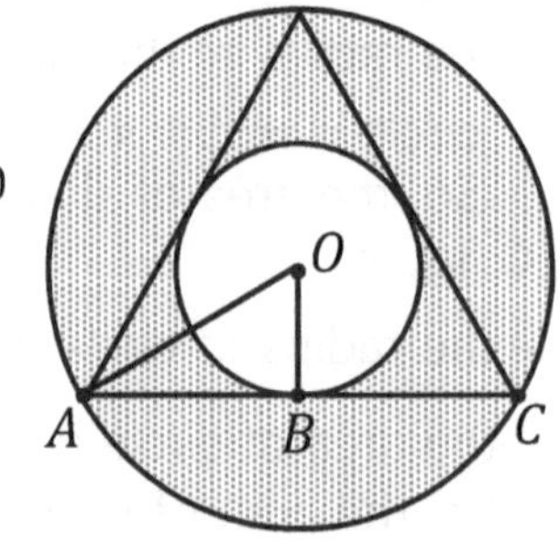

43. Answer $(\boldsymbol{B})$

Using the Pythagorean theorem.

① The area of two semicircles;

$$2 \times \left(\frac{1}{2} \times \pi(2)^2\right) = 4\pi$$

② The area of the larger circle;

$$R^2 = 2^2 + 2^2 = 8$$

$$\rightarrow \pi R^2 = 8\pi$$

So, the ratio is; $4\pi : 8\pi \rightarrow 1:2$

Thus, the correct answer is $\boldsymbol{B}$.

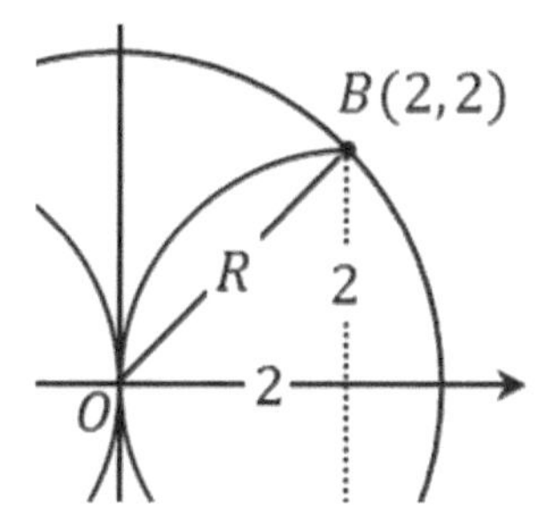

44. Answer $(\boldsymbol{E})$

The triangle is a right triangle.

Using formula of circle area, the side length of AB is;

$$18\pi = \frac{1}{2}\pi\left(\frac{AB}{2}\right)^2 \rightarrow \left(\frac{AB}{2}\right)^2 = 36 \rightarrow AB = 12$$

the side length of BC is;

$$50\pi = \frac{1}{2}\pi\left(\frac{BC}{2}\right)^2 \rightarrow \left(\frac{BC}{2}\right)^2 = 100 \rightarrow BC = 20.$$

The triangle is a right triangle, so using the Pythagorean theorem;

$$AC = \sqrt{20^2 - 12^2} = 14.$$

Therefore the area of triangle is;

$$Area = \frac{1}{2} \times 12 \times 14 = 84$$

Thus, the correct answer is $\boldsymbol{E}$.

45. Answer $(\boldsymbol{E})$

115 is an odd number.

115 is an odd number, even+odd=odd and 2 is only even prime number. So, $2 + 113 = 115$ and $113 - 2 = 111$.

Thus, the correct answer is $\boldsymbol{E}$.

46. **Answer** ($\boldsymbol{E}$)

The key is the difference is 2.

Because two-digit numbers have 5 for units digit are all composite numbers, R(Richard's score) and S(Samuel's score) are $11-13, 17-19, 41-43$, or $71-73$.
And P(Paul's score)$+S > 155$, $P+R < 155$, so the pair is $71-73$ and P is 83.

Thus, the correct answer is $\boldsymbol{E}$.

47. **Answer** ($\boldsymbol{C}$)

The product of prime is a composite.

The number is a product of prime numbers are greater than 60. The prime numbers that are greater than 60 are $61, 67, 71, \ldots$. So, the number is $61 \times 69 = 4209$.

Thus, the correct answer is $\boldsymbol{C}$.

48. **Answer** ($\boldsymbol{E}$)

Find a prime number.

At least half the numbers are too high, so his age should be less than 45. And two of the numbers are off by one, so his age is in between two numbers whose difference is 2. It could 37, 41, and 43 and since difference is 2 , the number is 43.$(\because 43-42=1 \; and \; 44-43=1)$.

Thus, the correct answer is $\boldsymbol{E}$.

49. **Answer** ($\boldsymbol{A}$)

The number is an even number.

The sum of two odd numbers is an even and the sum of even and odd is an odd number. There are only one even number, 2, in prime numbers so, $400004 = 400002 + 2$. 4002 is not prime number, therefore there are no cases.

Thus, the correct answer is $\boldsymbol{A}$.

50. **Answer** $(\boldsymbol{B})$

odd − even = odd.

Let A, B, and C be the hidden prime numbers and except for 2, all prime numbers are odd.

A	B	C
45	50	54

$45 - A = 50 - B$

So, A is an even prime number 2 and the difference is 43. So B is 7 and C is 11 and the sum $A + B + C = 2 + 7 + 11 = 20$.

Thus, the correct answer is $\boldsymbol{B}$.

51. **Answer** $(\boldsymbol{B})$

Do not prime factorization.

The prime factor of all even numbers is 2 and 2 is the smallest prime factor.

Thus, the correct answer is $\boldsymbol{B}$.

52. **Answer** $(\boldsymbol{B})$

The price of an apple is same.

Using prime factorization;

$$195 = 3 \times 5 \times 13, \qquad 165 = 3 \times 5 \times 11$$

And the price of an apple is more than \$10 and same so,

$$195 = 15 \times 13, \qquad 165 = 15 \times 11$$

Therefore mother (13)+farther (11)=26.

Thus, the correct answer is $\boldsymbol{B}$.

53. **Answer** $(\boldsymbol{C})$

Using the exponent rule.

A cubic number is a number of the form n^3 with n a positive integer. So,

$$4^{2019} = (4^{673})^3$$

Thus, the correct answer is $\boldsymbol{C}$.

54. **Answer** (***A***)

Using the prime factorization.

The prime factorization of 2025 is $3 \times 3 \times 3 \times 3 \times 5 \times 5 = 3^4 \times 5^2$.
The integer divisors of 2025 are $1, 3, 5, 9(= 3 \times 3), 15(= 3 \times 5), \ldots$,
so prime integer divisors are 3 and 5. $\rightarrow \quad 3 + 5 = 8$.

Thus, the correct answer is ***A***.

55. **Answer** (***D***)

Using the sum and difference formula.

From the sum and difference formula;

$$18^4 - 15^4 = (18^2)^2 - (15^2)^2 = (18^2 + 15^2)(18^2 - 15^2)$$
$$= (18^2 + 15^2)(18^2 - 15^2) = (18^2 + 15^2)(18 + 15)(18 - 15)$$
$$= (324 + 225)(33)(3) = 549 \times 33 \times 3 = 3 \times 3 \times 61 \times 3 \times 11 \times 3$$
$$= 3^4 \times 11 \times 61$$

Thus, the correct answer is ***D***.

56. **Answer** (***D***)

Using the prime factorization.

The prime factorization of 1960 is;

$$1960 = 2^3 \times 5^1 \times 7^2 = 2^3 \times 3^0 \times 5^1 \times 7^2$$

So, $2a - 3b + 5c - 7d = 2 \times 3 - 3 \times 0 + 5 \times 1 - 7 \times 2 = -3$
S

Thus, the correct answer is ***D***.

57. **Answer** (***E***)

Using the prime factorization.

Using the prime factorization of 2020;

$$2020 = 2^2 \times 5 \times 101$$

So, the sum is $2 + 5 + 101 = 108$.

Thus, the correct answer is ***E***.

58. **Answer** $(\boldsymbol{B})$

Using the prime factorization.

Find a common factor of 403 and 377 by using the prime factorization.

$$403 = 13 \times 31, \qquad 377 = 13 \times 29$$

So, the cost of a mechanical pencil is 13 and the numbers of students are 31 and 29.
Therefore, $31 - 29 = 2$.

Thus, the correct answer is $\boldsymbol{B}$.

59. **Answer** $(\boldsymbol{B})$

Using the prime factorization.

Using the prime factorization;

$$240 = 2^4 \times 3 \times 5$$

So, the product of 240 and A is a square means that

$$240 = 2^4 \times 3^1 \times 5^1 \times A = 2^4 \times 3^1 \times 5^1 \times 3^1 \times 5^1 = (2^2 \times 3 \times 5)^2$$

and the product of 240 and B is a cube means that

$$240 = 2^4 \times 3^1 \times 5^1 \times B = 2^4 \times 3^1 \times 5^1 \times 2^2 \times 3^2 \times 5^2$$
$$= (2^2 \times 3 \times 5)^3$$

Therefore, $A = 15$, $B = 900$, and $\frac{B}{A} = \frac{900}{15} = 60$.

Thus, the correct answer is $\boldsymbol{B}$.

60. **Answer** $(\boldsymbol{D})$

Using the prime factorization.

The prime factorization of 816 is;

$$816 = 2^4 \times 3 \times 17$$

So $3 + 17 = 20$

Thus, the correct answer is $\boldsymbol{D}$.

61. **Answer** ($\boldsymbol{C}$)

Using the power form.

① $2^a + 128 = 144 \rightarrow 2^a = 144 - 128 = 16 \rightarrow 2^a = 2^4 \rightarrow a = 4$
② $5^b - 38 = 587 \rightarrow 5^b = 587 + 38 = 625 \rightarrow 5^b = 5^4 \rightarrow b = 4$
③ $3^5 + 4^c = 1267 \rightarrow 4^c = 1267 - 3^5 = 1024 \rightarrow 4^c = 4^5 \rightarrow c = 5$
From ①, ②, and ③

$$a + b + c = 4 + 4 + 5 = 13$$

Thus, the correct answer is ***C***.

62. **Answer** ($\boldsymbol{D}$)

Using the exponent rule.

Using the exponent rule;
$2^{36} = (2^2)^{18} \rightarrow 4^{18} = 4^6 \times 4^{12}$, $\quad 5^{18} = 5^6 \times 5^{12}$
$10^{12} = 2^{12} \times 5^{12} \rightarrow 4^6 \times 5^{12}$
Since $4^6 < 5^6$ and $4^{12} < 5^{12}$, $2^{36} < 10^{12} < 5^{18}$

Thus, the correct answer is ***D***.

63. **Answer** ($\boldsymbol{D}$)

Finding the number of each bead

Counting the number of reaching each stair;
1st stair: 1 way
2nd stair: $1 \rightarrow 1,\ 2 \rightarrow 2$ ways
3rd stair: $1 \rightarrow 1 \rightarrow 1,\ 1 \rightarrow 2,\ 2 \rightarrow 1, 3 \rightarrow 4$ ways
On the 4th stair, he can get there all at once, even if he start from the 1st , 2nd , or 3rd stair. So,
4th stair: $1 + 2 + 4 = 7$ ways
And as same reason;
5th stair: $2 + 4 + 7 = 13$ ways (start from 2nd, 3rd, or 4th stair)
6th stair: $4 + 7 + 13 = 24$ ways (start from 3rd, 4th or 5th stair)
7th stair: $7 + 13 + 24 = 44$ ways
8th stair: $13 + 24 + 44 = 81$ ways

Thus, the correct answer is ***D***.

64. Answer (A)

There are 8 empty squares.

There are 2 ways to place the remaining two Xs

There are 2 ways to place the three Ys

There are 1 ways to place the three Ys
So, there are

$$2 \times 2 \times 1 = 4 \text{ ways}$$

Thus, the correct answer is $\boldsymbol{A}$.

		X
	X	
X		

		X
X		
	X	

Y		X
	X	Y
X	Y	

Y		X
X	Y	
	X	Y

Y	Z	X
Z	X	Y
X	Y	Z

Y	Z	X
X	Y	Z
Z	X	Y

65. Answer (B)

It is similar to shake hand.

The total number of connecting two point is calculated by combination formula;

$$10C2 = \frac{10!}{2!\,(10-2)!} = 45\, ways$$

The two points are one unit apart at 10 points around the perimeter of the rectangle, so there are 10 connecting.
Therefore, the probability is;

$$\frac{10}{45} = \frac{2}{9}$$

Thus, the correct answer is $\boldsymbol{B}$.

66. Answer (E)

The most unfortunate case is assumed..

The most unfortunate case is a case where the three colors continue to diverge, as like;
B, Y, G, B, Y, G, B, Y, G, B, Y, G, B, Y, G so 16th color is either B or Y or G.

Thus, the correct answer is $\boldsymbol{E}$.

67. Answer (*D*)

Finding the pattern.

The pattern is;
① the M is adjacent to 4 As
② each A is adjacent to 3 Ts
③ each T is adjacent to 2 Hs.
From ①, ②, and ③, the number of different paths is;

$$4 \times 3 \times 2 = 24 \text{ paths.}$$

Thus, the correct answer is ***D***.

68. Answer (*E*)

512 is a multiple of 8.

① the number of first race is; $\frac{512}{8} = 64$ races and 64 swimmers attends to the second race.
② the number of second race is; $\frac{64}{8} = 8$ races and 8 swimmers attends to the third race.
③ the number of third race is; $\frac{8}{8} = 1$ race.
From ①, ②, and ③ there are $64 + 8 + 1 = 73$ races.
Thus, the correct answer is ***E***.

69. Answer (*C*)

Seven blocks means the shortest path.

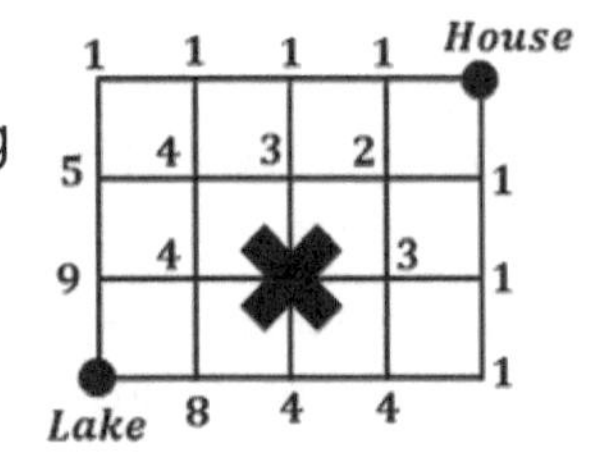

As shown as right,
the number of the minimum paths reaching the branch road of each block is counted and displayed.
So, the number of paths is;

$$9 + 8 = 17$$

Thus, the correct answer is ***C***.

70. **Answer** (C)

Drawing the given situation.

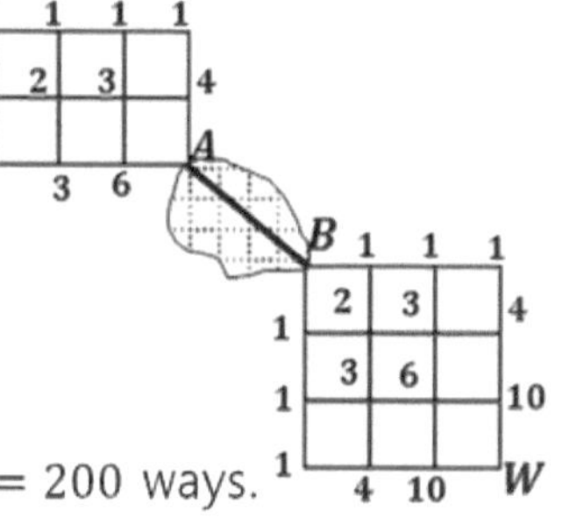

As shown as right,
the number of the minimum paths reaching the branch road of each block is counted and displayed.
① H to A is $6 + 4 = 10$ ways
② B to W is $10 + 10 = 20$ ways
So, the total number of ways is $10 \times 20 = 200$ ways.

Thus, the correct answer is C.

71. **Answer** (D)

Tree diagram is also good.

The possible outcomes are;

$$204, 240, 402, 420$$

So, the probability is $\frac{2}{4} = \frac{1}{2}$.
The

Thus, the correct answer is D.

72. **Answer** (D)

Counting the number of games on each round.

The number of games on each round is;
first round – 16 games, second – 8 games, third – 4 games, fourth – 2 games, fifth – 1 game,
So, $16 + 8 + 4 + 2 + 1 = 31$ games.

Thus, the correct answer is D.

73. **Answer** (C)

It is same array.

Let p be the total sum of 65 rows and r be the total sum of 30 columns. Being calculated array is same array so, the averages are;

$$average\ of\ row's\ sum = \frac{p}{65}, \qquad average\ of\ column's\ sum = \frac{r}{30}$$

and $p = r$.
The ratio is

$$\frac{\frac{p}{65}}{\frac{r}{30}} = \frac{\frac{1}{65}}{\frac{1}{30}} = \frac{30}{65} = \frac{6}{13}$$

Thus, the correct answer is $\boldsymbol{C}$.

74. **Answer** (E)

Using average formula.

Sum of seven scores is $5+8+2+4+7+5+4=35$. If the eighth score is x, the average is;

$$\frac{35+x}{8} = int \quad \rightarrow \quad \frac{(35+5)}{8} = \frac{40}{8} = 5 \quad \rightarrow \quad x = 5.$$

If the ninth score is y, the average is;

$$\frac{40+y}{9} = int \quad \rightarrow \quad \frac{(40+5)}{9} = \frac{45}{9} = 5 \quad \rightarrow \quad y = 5.$$

If the tenth score is z, the average is;

$$\frac{45+z}{10} = int \quad \rightarrow \quad \frac{(45+5)}{10} = \frac{50}{10} = 5 \quad \rightarrow \quad z = 5.$$

So, $x+y+z = 5+5+5 = 15$.

Thus, the correct answer is $\boldsymbol{E}$.

75. **Answer** (A)

Using the formula.

Let x be the fifth student's score;

$$\frac{65+75+80+90+x}{5} = 77 \quad \rightarrow \quad 310+x = 385 \quad \rightarrow \quad x = 75$$

Thus, the correct answer is $\boldsymbol{A}$.

76. **Answer** (A)

The highest possible score is 100.

His total sum of scores is $5 \times 90 = 450$ and the highest possible score on the fifth test is 100. Let x be the lowest possible score, then;

$$84 + 93 + 90 + x + 100 = 450 \quad \rightarrow \quad 367 + x = 450 \quad \rightarrow \quad x = 83$$

Thus, the correct answer is $\boldsymbol{A}$.

77. **Answer** $(\boldsymbol{B})$

Mean and average are same.

① Bessie

$70 + 50 + 80 + 50 + 30 = 280$ so, the mean is $\frac{280}{5} = 56.$

② Chloe

$40 + 80 + 70 + 60 + 20 = 270$ so, the mean is $\frac{280}{5} = 54.$

Thus, the correct answer is $\boldsymbol{B}$.

78. **Answer** $(\boldsymbol{C})$

Finding the total amount.

The total sum is;

$$75 + 150 + 175 + 125 + 200 = 725$$

So, the average sales per day is $\frac{725}{5} = 145.$

Thus, the correct answer is $\boldsymbol{C}$.

79. **Answer** $(\boldsymbol{B})$

Finding the total amount.

The mean age is;

$$\frac{7 \times 36 + 5 \times 42}{7 + 5} = \frac{462}{12} = 38.5$$

Thus, the correct answer is $\boldsymbol{B}$.

80. Answer $(\boldsymbol{D})$

Finding total hours.

Let f be the number of hours for the final week, then;

$$\frac{13+15+11+14+f}{5}=14 \quad \rightarrow \quad 53+f=70$$

So, $f=70-53=17$ hours for final week.

Thus, the correct answer is $\boldsymbol{D}$.

81. Answer $(\boldsymbol{C})$

Using average formula.

Let x be the total sum of ages;

$$\frac{x}{7}=35 \quad \rightarrow \quad x=245$$

So, the sum of ages of remaining people and average are;

$$\frac{245-11}{6}=\frac{234}{6}=39$$

Thus, the correct answer is $\boldsymbol{C}$.

82. Answer $(\boldsymbol{C})$

What is the total sum?

Let x be the sum of the six numbers, y be the sum of the first three numbers and z be the sum of the last three numbers;

$$\frac{x}{6}=42 \ \rightarrow \ x=252, \qquad \frac{y}{3}=36 \ \rightarrow \ y=108$$

So, $252=108+z \ \rightarrow \ z=144$ and $\frac{144}{3}=48$

Thus, the correct answer is $\boldsymbol{C}$.

83. Answer $(\boldsymbol{A})$

Find the difference total sum.

Maggie' sum is $89 \times 5=445$, so Owen's sum and average are;

$$445-8+12+11-15-15=430 \ \rightarrow \ \frac{430}{5}=86$$

And the difference is 3.

Thus, the correct answer is $\boldsymbol{A}$.

84. **Answer** $(\boldsymbol{C})$

Finding the smallest sum.

The smallest six distinct positive odd integers are $1, 3, 5, 7, 9, 11$, so the average is;

$$\frac{1+3+5+7+9+11}{6} = \frac{36}{6} = 6$$

Thus, the correct answer is $\boldsymbol{C}$.

85. **Answer** $(\boldsymbol{E})$

Converting time to minutes.

Let x be the tenth day jogging time and the average is;

$$\frac{5 \times 80 + 4 \times 100 + x}{10} = 90 \quad \rightarrow \quad 800 + x = 900 \quad \rightarrow \quad x = 100$$

So, the jogging time of tenth day is 100 minutes$=1$ hour 40 minutes.

Thus, the correct answer is $\boldsymbol{E}$.

PRIMUS INTER PARES
K·DEAN

Made in United States
Troutdale, OR
11/07/2024